Representation in Cognitive Science

Representation in Cognitive Science

Nicholas Shea

OXFORD
UNIVERSITY PRESS

OXFORD
UNIVERSITY PRESS

Great Clarendon Street, Oxford, OX2 6DP,
United Kingdom

Oxford University Press is a department of the University of Oxford.
It furthers the University's objective of excellence in research, scholarship,
and education by publishing worldwide. Oxford is a registered trade mark of
Oxford University Press in the UK and in certain other countries

First Edition published in 2018

Impression: 2

Published in the United States of America by Oxford University Press
198 Madison Avenue, New York, NY 10016, United States of America

British Library Cataloguing in Publication Data
Data available

Library of Congress Control Number: 2018939590

ISBN 978–0–19–881288–3

Printed and bound by
CPI Group (UK) Ltd, Croydon, CR0 4YY

To Ellie

*Huffing and puffing with correlation and function might give us a good account of subdoxastic aboutness . . . [but it is unlikely to work for the content of doxastic states.]**

Martin Davies (pers. comm.), developed in Davies (2005)

* Not that the antecedent is easy. And even for the subpersonal case we may have to puff on a few more ingredients. But I too am optimistic that we can get a good account. This book aims to show how.

Preface

The book is in three parts: introduction, exposition, and defence. Part I is introductory and light on argument. Chapter 1 is about others' views. Chapter 2 is the framework for my own view. I don't rehearse well-known arguments but simply gesture at the literature. The aim is to demarcate the problem and motivate my own approach. Part II changes gear, into more standard philosophical mode. It aims to state my positive view precisely and to test it against a series of case studies from cognitive science. Part III engages more carefully with the existing literature, showing that the account developed in Part II can deal with important arguments made by previous researchers, and arguing that the framework put forward in Part I has been vindicated.

There is a paragraph-by-paragraph summary at the end of the book. Readers who want to go straight to a particular issue may find this a useful guide. It replaces the chapter summaries often found at the end of each chapter of a monograph. The bibliography lists the pages where I discuss each reference, so acts as a fine-grained index to particular issues. There is also the usual keyword index at the end.

Contents

PART I

1

Introduction

1.1 A Foundational Question

The mind holds many mysteries. Thinking used to be one of them. Staring idly out of the window, a chain of thought runs through my mind. Concentrating hard to solve a problem, I reason my way through a series of ideas until I find an answer (if I'm lucky). Having thoughts running through our minds is one of the most obvious aspects of the lived human experience. It seems central to the way we behave, especially in the cases we care most about. But what are thoughts and what is this process we call thinking? That was once as mysterious as the movement of the heavens or the nature of life itself.

New technology can fundamentally change our understanding of what is possible and what mysterious. For Descartes, mechanical automata were a revelation. These fairground curiosities moved in ways that looked animate, uncannily like the movements of animals and even people. A capacity that had previously been linked inextricably to a fundamental life force, or to the soul, could now be seen as purely mechanical. Descartes famously argued that this could only go so far. Mechanism would not explain consciousness, nor the capacity for free will. Nor, he thought, could mechanism explain linguistic competence. It was inconceivable that a machine could produce different arrangements of words so as to give an appropriately grammatical answer to questions asked of it.[1] Consciousness and free will remain baffling. But another machine has made what was inconceivable to Descartes an everyday reality to us.

Computers produce appropriately arranged strings of words—Google even annoyingly finishes half-typed sentences—in ways that at least respect the meaning of the words they churn out. Until quite recently a 'computer' was a person who did calculations. Now we know that calculations can be done mechanically. Babbage,

[1] Descartes (1637/1988, p. 44: AT VI 56: CSM I 140), quoted by Stoljar (2001, pp. 405–6).

Lovelace, and others in the nineteenth century saw the possibility of general-purpose mechanical computation, but it wasn't until the valve-based, then transistor-based computers of the twentieth century that it became apparent just how powerful this idea was.[2]

This remarkable insight can also answer our question about thinking: the answer is that thinking is the processing of mental representations. We're familiar with words and symbols as representations, from marks made on a wet clay tablet to texts appearing on the latest electronic tablet: they are items with meaning.[3] A written sentence is a representation that takes the form of ink marks on paper: 'roses are red'. It also has meaning—it is about flowers and their colour. Mental representations are similar: I believe that today is Tuesday, see that there is an apple in the bowl, hope that the sun will come out, and think about an exciting mountain climb. These thoughts are all mental representations. The core is the same as with words and symbols. Mental representations are physical things with meaning. A train of thought is a series of mental representations. That is the so-called 'representational theory of mind'.

I say the representational theory of mind is 'an' answer to our question about thinking, not 'the' answer, because not everyone agrees it is a good idea to appeal to mental representations. Granted, doing physical manipulations on things that have meaning is a great idea. We count on our fingers to add up. We manipulate symbols on the page to arrive at a mathematical proof. The physical stuff being manipulated can take many forms. Babbage's difference engine uses gears and cogs to do long multiplication (see Figure 1.1). And now our amazingly powerful computers can do this kind of thing at inhuman speed on an astonishing scale. They manipulate voltage levels not fingers and can do a lot more than work out how many eggs will be left after breakfast. But they too work by performing physical manipulations on representations. The only trouble with carrying this over to the case of thinking is that we're not really sure how mental representations get their meaning.

For myself, I do think that the idea of mental representation is the answer to the mystery of thinking. There is very good reason to believe that thinking is the processing of meaningful physical entities, mental representations. That insight is one of the most important discoveries of the twentieth century—it may turn out to be *the* most important. But I have to admit that the question of meaning is a little problem in the foundations. We've done well on the 'processing' bit but we're still a bit iffy about the 'meaningful' bit. We know what processing of physical particulars is, and how processing can *respect* the meaning of symbols. For example, we can make a machine whose manipulations obey logical rules and so preserve truth. But we don't yet have a clear idea of how representations could *get* meanings, when the meaning does not derive from the understanding of an external interpreter.

[2] Developments in logic, notably by Frege, were of course an important intermediate step, on which Turing, von Neumann, and others built in designing computing machines.

[3] We'll have to stretch the point for some of my son's texts.

Figure 1.1 Babbage's difference engine uses cogs and gears to perform physical manipulations on representations of numbers. It is used to multiply large numbers together. The components are representations of numbers and the physical manipulations make sense in the light of those contents—they multiply the numbers (using the method of differences).

So, the question remains: how do mental states[4] manage to be about things in the external world? That mental representations are about things in the world, although utterly commonplace, is deeply puzzling. How do they get their *aboutness*? The physical and biological sciences offer no model of how naturalistically respectable properties could be like that. This is an undoubted lacuna in our understanding, a void hidden away in the foundations of the cognitive sciences. We behave in ways that are suited to our environment. We do so by representing the world and processing those representations in rational ways—at least, there is strong evidence that we do in very many cases. Mental representations represent objects and properties in the world: the

[4] I use 'mental' broadly to cover all aspects of an agent's psychology, including unconscious and/or low-level information processing; and 'state' loosely, so as to include dynamic states, i.e. events and processes. 'Mental state' is a convenient shorthand for entities of all kinds that are psychological and bear content.

shape of a fruit, the movement of an animal, the expression on a face. I work out how much pasta to cook by thinking about how many people there will be for dinner and how much each they will eat. 'Content' is a useful shorthand for the objects, properties and conditions that a representation refers to or is about. So, the content of one of my thoughts about dinner is: *each person needs 150g of pasta*.

What then is the link between a mental representation and its content? The content of a representation must depend somehow on the way it is produced in response to input, the way it interacts with other representations, and the behaviour that results. How do those processes link a mental representation with the external objects and properties it refers to? How does the thought in my head connect up with quantities of pasta? In short: what determines the content of a mental representation? That is the 'content question'. Surprisingly, there is no agreed answer.

This little foundational worry hasn't stopped the cognitive sciences getting on and using the idea of mental representation to great effect. Representational explanation is the central resource of scientific psychology. Many kinds of behaviour have been convincingly explained in terms of the internal algorithms or heuristics by which they are generated. Ever since the 'cognitive revolution' gave the behavioural sciences the idea of mental representation, one phenomenon after another has succumbed to representational explanation, from the trajectories of the limbs when reaching to pick up an object, to parsing the grammar of a sentence. The recent successes of cognitive neuroscience depend on the same insight, while also telling us how representations are realized in the brain, a kind of understanding until recently thought to be fanciful. Figure 1.2 shows a typical example. The details of this experiment need not detain us for now (detailed case studies come in Part II). Just focus on the explanatory scheme. There is a set of interconnected brain areas, plus a computation being performed by those brain areas (sketched in the lower half of panel (a)). Together that tells us how participants in the experiment manage to perform their task (inset). So, although we lack a theory of it, there is little reason to doubt the existence of representational content. We're in the position of the academic in the cartoon musing, 'Well it works in practice, Bob, but I'm not sure it's really gonna work in theory.'

The lack of an answer to the content question does arouse suspicion that mental representation is a dubious concept. Some want to eliminate the notion of representational content from our theorizing entirely, perhaps replacing it with a purely neural account of behavioural mechanisms. If that were right, it would radically revise our conception of ourselves as reason-guided agents since reasons are mental contents. That conception runs deep in the humanities and social sciences, not to mention ordinary life. But even neuroscientists should want to hold onto the idea of representation, because their explanations would be seriously impoverished without it. Even when the causes of behaviour can be picked out in neural terms, our understanding of why *that* pattern of neural activity produces *this* kind of behaviour depends crucially on neural activity being about things in the organism's environment. Figure 1.2 doesn't just show neural areas, but also how the activity of those areas should be understood as

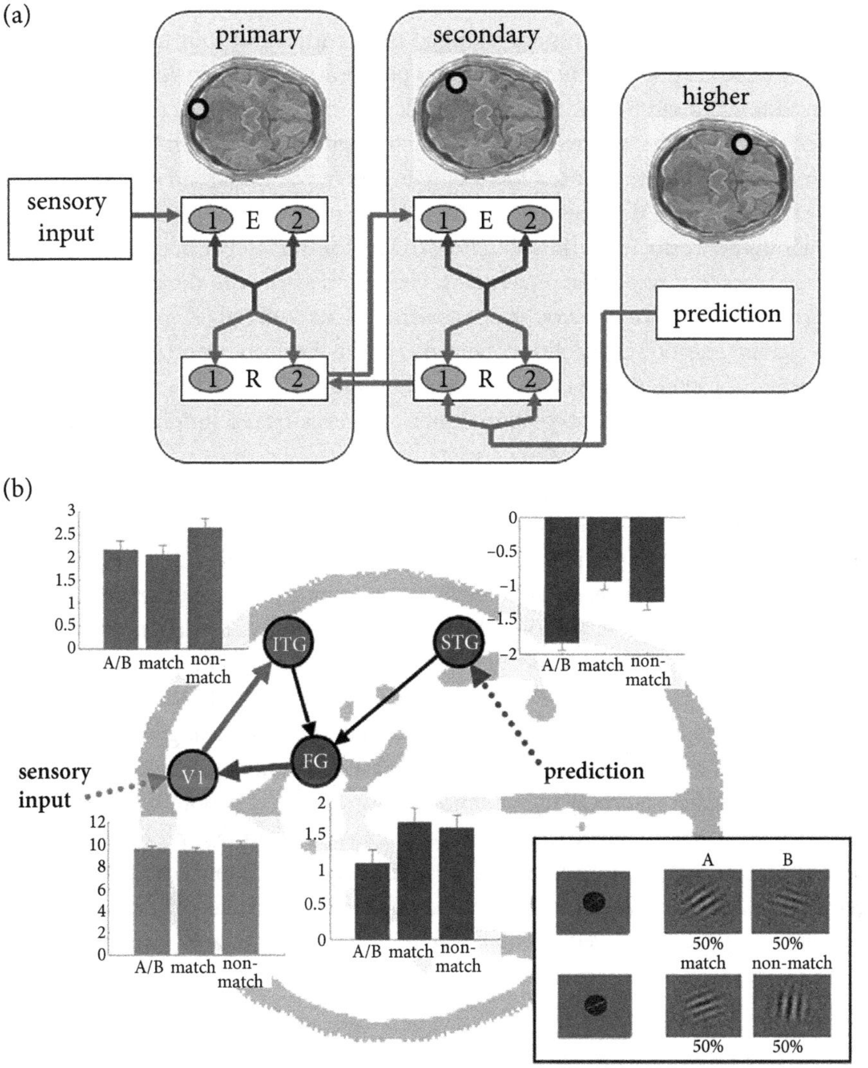

Current Opinion in Neurobiology

Figure 1.2 A figure that illustrates the explanatory scheme typical of cognitive neuroscience (from Rushworth et al. 2009). There is a computation (sketched in the lower half of panel (a)), implemented in some interacting brain areas, so as to perform a behavioural task (inset). The details are not important for present purposes.

representing things about the stimuli presented to the people doing a task. The content of a neural representation makes an explanatory connection with distal features of the agent's environment, features that the agent reacts to and then acts on.

One aspect of the problem is consciousness. I want to set that aside. Consciousness raises a host of additional difficulties. Furthermore, there are cases of thinking and reasoning, or something very like it, that go on in the absence of consciousness. Just to walk along the street, your eyes are taking in information and your mind is tracking the direction of the path and the movement of people around you. Information is being processed to work out how to adjust your gait instant by instant, so that you stay on your feet and avoid the inconvenience of colliding with the person in front engrossed in their smartphone. I say those processes going on in you are a kind of reasoning, or are like familiar thought processes, because they too proceed through a series of states, states about the world, in working out how to act. They involve processing representations in ways that respect their contents. Getting to grips with the content of non-conscious representations is enough of a challenge in its own right.[5]

The content question is widely recognized as one of the deepest and most significant problems in the philosophy of mind, a central question about the mind's place in nature. It is not just of interest to philosophers, however. Its resolution is also important for the cognitive sciences. Many disputes in psychology concern which properties are being represented in a particular case. Does the mirror neuron subsystem represent other agents' goals or merely action patterns (Gallese et al. 1996)? Does the brain code for scalar quantities or probability distributions (Pouget et al. 2003)? Do infants represent other agents' belief states or are they just keeping track of behaviour (Apperly and Butterfill 2009)? Often such disputes go beyond disagreements about what causal sensitivities and behavioural dispositions the organism has. Theorists disagree about what is being represented in the light of those facts. What researchers lack is a soundly based theory of content which tells us what is being represented, given established facts about what an organism or other system responds to and how it behaves.

This chapter offers a breezy introduction to the content question for non-experts. I gesture at existing literature to help demarcate the problem, but proper arguments will come later (Parts II and III). So that I can move quickly to presenting the positive account (Chapter 2 onwards), this chapter is more presupposition than demonstration. It should bring the problem of mental content into view for those unfamiliar with it, but it only offers my own particular take on the problem.

1.2 Homing In on the Problem

The problem of mental content in its modern incarnation goes back to Franz Brentano in the nineteenth century. Brentano identified aboutness or 'intentionality'[6] as being a peculiar feature of thoughts (Brentano 1874/1995). Thoughts can be about objects and properties that are not present to the thinker (the apple in my rucksack), are distant in time and space (a mountain in Tibet), are hypothetical or may only lie far in the future (the explosion of the sun), or are entirely imaginary (Harry Potter). How can mental

[5] Roughly, I'm setting aside beliefs and desires (doxastic states) and conscious states—see §2.1. I use 'subpersonal' as a label for mental representations that don't have these complicating features.

[6] This is a technical term—it's not about intentions.

states reach out and be about such things? Indeed, how do beliefs and perceptual states manage to be about an object that is right in front of the thinker (the pen on my desk), when the object is out there, and the representations are inside the thinker?

We could ask the same question about the intentionality of words and natural language sentences: how do they get their meaning? An obvious answer is: from the thoughts of the language users.[7] The meaning of a word is plausibly dependent on what people usually take it to mean: 'cat' is about cats because the word makes people *think* about cats. That kind of story cannot then be told about mental representations, on pain of regress. In order to start somewhere we start with the idea that at least some mental representations have underived intentionality. If we can't make sense of underived intentionality somewhere in the picture—of where meaning ultimately comes from—then the whole framework of explaining behaviour in terms of what people perceive and think is resting on questionable foundations. The most fruitful idea in the cognitive sciences, the idea of mental representation, which we thought we understood, would turn out to be deeply mysterious, as difficult as free will or as consciousness itself.

When asked about the content of a familiar mental representation like a concept, one common reaction is to talk about other mental states it is associated with. Why is my concept DOG about *dogs*?[8] Because it brings to mind images of dogs, the sound of dogs barking, the feel of their fur and their distinctive doggy smell. We'll come back to these kinds of theories of content in the next section, but for now I want to point out that this answer also just pushes back the question: where do mental images get their contents? In virtue of what do they represent the visual features, sounds, tactile properties and odours that they do? Underived intentionality must enter the picture somewhere.

The task then is to give an account of how at least some mental representations have contents that do not derive from the contents of other representations. What we are after is an account of what determines the content of a mental representation, determination in the metaphysical sense (what makes it the case that a representation has the content it does?) not the epistemic sense (how can we tell what the content of mental representation is?). An object-level semantic theory gives the content of mental representations in a domain (e.g. tells us that cognitive maps refer to spatial locations). Many information-processing accounts of behaviour offer a semantic theory in this sense. They assign correctness conditions and satisfaction conditions to a series of mental representations and go on to say how those representations are involved in generating intelligent behaviour. Our question is a meta-level question

[7] Another tenable view is that sentences have underived intentionality. For beliefs and desires, the claim that their content derives from the content of natural language sentences has to be taken seriously. But here I set aside the problem of belief/desire content (§2.1) to focus on simpler cases in cognitive science.

[8] I use small caps to name concepts; and italics when giving the content of a representation (whether a subpropositional constituent, as here, or a full correctness condition or satisfaction condition). I also use italics when introducing a term by using (rather than mentioning) it.

about these theories: in virtue of what do those representations have those contents (if indeed they do)? For example, in virtue of what are cognitive maps about locations in the world? Our task then is to formulate a meta-semantic theory of mental representation.

It is popular to distinguish the question of what makes a state a representation from the question of what determines its content (Ramsey 2007). I don't make that distinction. To understand representational content, we need an answer to both questions. Accordingly, the accounts I put forward say what makes it the case, both that some state is a representation, and that it is a representation with a certain content.

We advert to the content of representations in order to explain behaviour. To explain how you swerve to avoid an oncoming smartphone zombie, the psychologist points to mental processes that are tracking the trajectory of people around you. A theory of content can illuminate how representations play this kind of explanatory role. One central explanatory practice is to use correct representations to explain successful behaviour. That assumption is more obviously at work in its corollary: misrepresentation explains failure. Because she misperceived the ground, she stumbled. Because he thought it was not yet eight o'clock, he missed the train. Misrepresentation explains why behaviour fails to go smoothly or fails to meet an agent's needs or goals. When things go badly for an agent, we can often pin the blame on an incorrect representation. We can also often explain the way they do behave; for instance, misrepresenting the time by fifteen minutes explains why he arrived on the platform nearly fifteen minutes after the train left.

Misrepresentation is one of the most puzzling aspects of representational content. A mental representation is an internal physical particular. It could be a complex pattern of neural activity. Cells firing in the hippocampus tell the rat where it is in space so it can work out how to get to some food at another location. If the cell firing misrepresents its current location, the rat will set off in the wrong direction and fail to get to the food. Whether a representation is correct or incorrect depends on factors outside the organism, which seem to make no difference to how the representation is processed within the organism (e.g. to how activity of some neurons causes activity of others). Yet its truth or falsity, correctness or incorrectness, is supposed to make a crucial explanatory difference. The capacity to misrepresent, then, is clearly a key part of what makes representational content a special kind of property, a target of philosophical interest. Any good theory of content must be able to account for misrepresentation.

A theory of content need not faithfully recapitulate the contents relied on in psychological or everyday explanations of behaviour. It may be revisionary in some respects, sometimes implying that what is actually represented is different than previously thought. Indeed, a theory of content can, as I suggested, help arbitrate disputes between different proposed content assignments.[9] However, it should deliver reasonably determinate contents. A theory of content needs to be applicable in concrete cases.

[9] E.g. whether infants are tracking others' mental states or just their behaviour.

For example, reinforcement learning based on the dopamine subsystem explains the behaviour elicited in a wide range of psychological experiments. We can predict what people will choose if we know how they have been rewarded for past choices. Plugging in facts about what is going on in a concrete case, a theory of content should output correctness conditions and/or satisfaction conditions for the representations involved. The determinacy of those conditions needs to be commensurate with the way correct and incorrect representation explains successful and unsuccessful behaviour in the case in question. A theory of content would obviously be hopeless if it implied that every state in a system represents every object and property the system interacts with. Delivering appropriately determinate contents is an adequacy condition on theories of content.

The problem of determinacy has several more specific incarnations. One asks about causal chains leading up to a representation. When I see a dog and think about it, is my thought about the distal object or about the pattern of light on my retina? More pointedly, can a theory of content distinguish between these, so that it implies that some mental representations have distal contents, while others represent more proximal matters of fact? A second problem is that the objects we think about exemplify a whole host of properties at once: the dog is a member of the kind *dog*, is brown and furry, is a medium-sized pliable physical object, and so on. The *qua* problem asks which of these properties is represented. Finally, for any candidate contents, we can ask about their disjunction. A theory may not select between *that is a dog* and *that is a brown, furry physical object* but instead imply that a state represents *that is a dog or that is a brown, furry physical object*. Rather than misrepresenting an odd-looking fox as a dog, I would end up correctly representing it as a brown furry object. If every condition in which this representation happens to be produced were included, encompassing things like shaggy sheep seen from an odd angle in poor light, then the representation would never end up being false. Every condition would be found somewhere in the long disjunction. We would lose the capacity to misrepresent. For that reason, the adequacy condition that a theory of content should imply or explain the capacity for misrepresentation is sometimes called the 'disjunction problem'. The *qua* problem, the disjunction problem, and the problem of proximal/distal contents are all different guises of the overall problem of determinacy.

Since we are puzzled about how there could be representational contents, an account of content should show how content arises out of something we find less mysterious. An account in terms of the phenomenal character of conscious experience, to take one example, would fail in this respect.[10] Standardly, naturalistic approaches offer accounts of content that are non-semantic, non-mental, and non-normative. I am aiming for an account that is naturalistic in that sense. Of course, it may turn out that there

[10] That is not in itself an argument against such theories—it could turn out that intentionality can only be properly accounted for in phenomenal terms—but it is a motivation to see if a non-phenomenal theory can work.

is no such account to be had. But in the absence of a compelling a priori argument that no naturalistic account of mental representation is possible, the tenability of the naturalistic approach can only properly be judged by the success or failure of the attempt. The project of naturalizing content must be judged by its fruits.

1.3 Existing Approaches

This section looks briefly at existing approaches to content determination. I won't attempt to make a case against these approaches. Those arguments have already been widely canvassed. My aim is to introduce the main obstacles these theories have faced, since these are the issues we will have to grapple with when assessing the accounts of content I put forward in the rest of the book. Although the theories below were advanced to account for the content of beliefs, desires, and conscious states, the same issues arise when they are applied to neural representations and the other cases from cognitive science which form the focus of this book.

One obvious starting point is information in the weak sense of correlation.[11] Correlational information arises whenever the states of items correlate, so that item X's being in one state (smoke is coming from the windows) raises the probability that item Y is in another state (there is a fire in the house). A certain pattern of neural firing raises the probability that there is a vertical edge in the centre of the visual field. If the pattern of firing is a neural representation, then its content may depend on the fact that this pattern of activity makes it likely that there is a vertical edge in front of the person.

Information theory has given us a rich understanding of the properties of information in this correlational sense (Cover and Thomas 2006). However, for reasons that have been widely discussed, representational content is clearly not the same thing as correlational information. The 'information' of information-processing psychology is a matter of correctness conditions or satisfaction conditions, something richer than the correlational information of information theory. Sophisticated treatments use the tools of information theory to construct a theory of content which respects this distinction (Usher 2001, Eliasmith 2013). However, the underlying liberality of correlational information continues to make life difficult. Any representation carries correlational information about very many conditions at once, so correlation does not on its own deliver determinate contents. Some correlations may be quite weak, and it is not at all plausible that the content of a representation is the thing it correlates with most strongly.[12] A weak correlation that only slightly raises the

[11] Shannon (1948) developed a formal treatment of correlational information—as a theory of communication, rather than meaning—which forms the foundation of (mathematical) information theory. Dretske (1981) applied information theory to the problem of mental content.

[12] Usually the strongest correlational information carried by a neural representation concerns other neural representations, its proximal causes, and its effects. The same point is made in the existing literature about beliefs. My belief that there is milk in the fridge strongly raises the probability that I have been thinking about food, only rather less strongly that there actually is milk in the fridge.

chance that there is a predator nearby may be relied on for that information when the outcome is a matter of life or death. Representations often concern distal facts, like the presence of a certain foodstuff, even though they correlate more strongly with a proximate sensory signal. Furthermore, a disjunction of conditions is always more likely than conditions taken individually: for example, an object might be an eagle, but it is more likely that it is an eagle *or* a crow. So, content-as-probability-raising faces a particularly acute form of the disjunction problem. Correlational information may well be an ingredient in a theory of content (Chapter 4), but even the sophisticated tools of mathematical information theory are not enough, without other ingredients, to capture the core explanatory difference between correct representation and misrepresentation.

Another tactic looks to relations between representations to fix content. We saw the idea earlier that the concept DOG gets its meaning from the inferences by which it is formed, such as from perceiving a brown furry object, and perhaps also from conclusions it generates, such as inferring that this thing might bite me (Block 1986). Patterns of inferences are plausibly what changes when a child acquires a new mathematical concept (Carey 2009). They are also the focus of recent Bayesian models of causal learning (Gopnik and Wellman 2012, Danks 2014). Moving beyond beliefs to neural representations, dispositions to make inferences—that is, to transition between representations—could fix content here too. If all inferences are relevant, holism threatens (Fodor and Lepore 1992): any change anywhere in the thinker's total representational scheme would change the content of all of their representations. There have been attempts to identify, for various concepts, a privileged set of dispositions that are constitutive of content (e.g. Peacocke 1992). However, it has proven difficult to identify sets of inferences that can do the job: that are necessary for possessing the concept, plausibly shared by most users of the concept, and sufficiently detailed to be individuative—that is, to distinguish between different concepts. For these reasons, inferential role semantics has not had much success in naturalizing content, except perhaps for the logical constants. The same concerns carry over when we move from beliefs to other representations relied on in the cognitive sciences.[13]

Relations amongst representations may be important for another reason. They endow a system of representations with a structure, which may mirror a structure in the world. For example, spatial relations between symbols on a map mirror spatial relations between places on the ground; and that seems to be important to the way maps represent. In the same way, Paul Churchland has argued that the similarity structure on a set of mental representations of human faces determines that they pick out certain

[13] Concepts (the constituents of beliefs) are usually thought to have neo-Fregean sense, as well as referential content (content that contributes to truth conditions). We may well have to appeal to inferential relations between concepts to account for differences in sense between co-referential concepts, and/or to vehicle properties (Millikan 2000, Sainsbury and Tye 2007, Recanati 2012). This book does not deal with concepts. I leave aside the issue of whether we need to appeal to neo-Fregean senses in addition to vehicle properties and referential contents.

individuals (Churchland 1998, 2012). Taken on its own the correspondence idea produces an implausibly liberal theory of representation (Cummins 1989, Godfrey-Smith 1994a, Shea 2013c). As we will see, however, structural correspondence is another plausible ingredient in a theory of content (Chapter 5).

Another group of theories are ascriptionist: they ascribe mental states to the whole person based on how she behaves, but don't commit to mental representations being physical particulars. This eschews what I described above as the core insight of the representational theory of mind (RTM). I discuss ascriptionism here because it remains a viable alternative to RTM (Williams 2016, 2018), especially for beliefs and desires,[14] so it will be important to be clear about the explanatory advantages that flow from the commitment to representations as physical particulars, when that commitment is warranted (see §2.5). Donald Davidson's version of the view derives from rational decision theory (Davidson 1974a, 1974b). The choices of an agent who obeys some minimal conditions of rationality can be modelled as if the agent has an ordered set of preferences about states of the world combined with a set of probabilistic beliefs about the likelihood of world states and the chance that actions she can perform would bring about other world states. According to Davidson, for an agent to be interpretable in this way is a key part of what it is to have beliefs and desires, to be a representer.

Daniel Dennett's intentional stance is in the same family of views (Dennett 1981). He emphasizes that there is nothing irrealist about this approach. People and other agents are tied up in patterns of interaction with the world that we can predict and explain from the intentional stance—that is, by treating them as having beliefs and desires. We could not do so if these interactions were described in purely physical terms, for example in terms of energy transduced by sensory receptors, producing neural states, which generate movements of the limbs. I can arrange to meet a colleague at a café in far-away Canberra three months hence. The intentional stance allows me to predict where they will be at 10 a.m. on 1 July in a way that would be impossible in practice by just keeping track of their moment-by-moment physical interactions with their environment. Even if it were not intractable, although a purely physical description would tell us, in physical terms, what is going to happen instant-by-instant, it would miss out on real patterns that exist in the behaviour (Dennett 1991). Those real patterns only come into view when we take the intentional stance, but the patterns are there irrespective of whether we recognize them (see §2.3 and §8.2b). The ontology of patterns means there is an observer-independent fact of the matter about which systems are interpretable from the intentional stance.

Dennett's ascriptionism is a tenable and substantial naturalistic account of representational content.[15] In this sense we already have a good theory of content. It is realist

[14] Neither Davidson nor Dennett claimed that their ascriptionism could be extended to the neural representations that are characteristic of the case studies we consider here.

[15] He advances it not for neural representations but as an account of belief-desire content. Davidson's account is not naturalistic in our sense. He argues that it is not possible to give an account of content in non-normative terms.

about what it takes to be a representer. However, I will reserve the term 'realism' for accounts that are committed to there being real *vehicles* of content: individuable physical particulars that bear contents and whose causal interactions explain behaviour. As I have been describing the problem of mental content, realism about vehicles is a core part of what it takes to be a mental representation. There are many cases where we have good evidence for realism about mental representations. We have already seen some examples in passing; subsequent chapters go into detail about many more. Where there are vehicles, representational explanation can explain more (§2.5). So, my task is to formulate an account of content applicable to cases where we have good reason to be realist about mental representations.

1.4 Teleosemantics

Teleosemantics is the final stop on our whistle-stop tour of the problems faced by existing theories of content. We will look at this family of views in slightly more detail since teleosemantics is the closest precursor to the accounts presented in this book. Teleosemantic views are those that give etiological functions a content-fixing role. This does not exclude there also being a role for correlational information, inferential role, or structural correspondence. A second commitment of the teleosemantic views of Ruth Millikan and David Papineau is a central role for a representation *consumer*: an identifiable subsystem that takes representations as input and generates outputs in response (Millikan 1984, 1989, Papineau 1987, 2016).

Peter Godfrey-Smith calls a commitment to representation consumers the 'basic representationalist model' (Godfrey-Smith 2006). It goes beyond the standard representational theory of mind (RTM), that is the commitment to representations as causally interacting particulars. The central idea of the basic representationalist model is that a representation is a stand-in that is relied on by a consumer to allow it to deal with some external state of affairs (see Figure 1.3). The consumer uses the state X as a guide to something else Y that it does not have access to directly. The idea is not that the consumer reads or interprets the representation, but simply that it reacts to

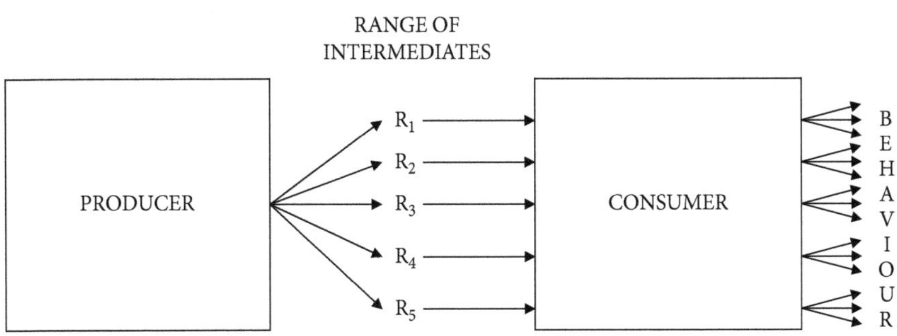

Figure 1.3 The basic representationalist model.

an intermediate state in a particular way. For example, 'consumer' honeybees observe dances of incoming bees as a guide to where nectar is located. In that case the representation is out in the open. In most psychological cases the representation is internal and the consumer is a subsystem within the organism. Ruth Millikan's teleosemantic theory is also committed to there being an identifiable representation producer.

Informational approaches to content direct our attention to the way a representation is produced. Conditions in the world cause a representation to be tokened;[16] the representation having been produced raises the probability that those conditions obtain. Consumer-based views invert the picture. Downstream effects on a consumer both constitute states as representations and fix their contents. What a representation means depends on how it is used downstream, on what it is taken to mean by a consumer of the representation. If the organism is relying on R as a stand-in, the way the consumer behaves in response to R will betray what it is taking R to mean.[17] Papineau's version of this idea targets beliefs and desires in the first instance (Papineau 1987, but see Papineau 2003). To see what a person believes, see how they act to satisfy their desires. So, the content of a belief is roughly the condition under which the behaviour it prompts would satisfy the consumer's desires. Sitting at my laptop, an electronic sound prompts an internal state R in me, which causes me to click on an icon to open my inbox. Given my desire to read messages sent to me, state R has the content *there is a new email message for you*. The representation R is separate from the consumer subsystem. The content of the representation derives from the way the consumer reacts to R.

For Millikan, the content of a representation is the condition under which behaviour of the consumer, prompted by the representation, will be successful (Millikan 1984). The distinctive contribution of teleosemantics is to understand success in evolutionary terms. The behaviour of the consumer subsystem has evolutionary functions. Success is a matter of performing those evolutionary functions so as to promote survival and reproduction. The success conditions for a behaviour are the conditions which obtained when behaviour of that type was selected. They are conditions which explain why behaviour of that type led systematically to survival and reproduction.

Consider the way honeybees communicate about where to forage (Figure 1.4). Incoming bees that have found a source of nectar are producers. They perform a dance that indicates the location of the nectar. The direction of the dance correlates with the direction of nectar and the time spent waggling correlates with distance. Outgoing bees are consumers. They condition their foraging behaviour on the dance. The dance acts as a stand-in for the location of nectar, something the outgoing bees have no direct

[16] A representation is *tokened* when an instance of it is realized. E.g. the rat has an array of place cells that represent locations. One of these representations is tokened when a place cell is active.

[17] The same idea is in Braithwaite (1933): I believe that p means that, under relevant external circumstances, relative to my needs, I will behave in a manner appropriate to p. Braithwaite also anticipated a naturalistic treatment of what it is for an action to be appropriate to a person's needs: 'satisfaction of these needs is something of which I do not despair of a naturalistic explanation'. Success semantics has the same structure (Whyte 1990).

Figure 1.4 The dance of the honeybee indicates the location of a source of nectar.

access to. The behaviour of the consumer is to fly off in a direction and for a distance that corresponds to the dance they have observed and then to start foraging at that location. This pattern of behaviour is very likely to be the result of natural selection on ancestor colonies of bees. For each type of dance there is an associated specific condition; for example, two seconds of vertical waggling might correspond to there being nectar 400 metres away in the direction of the sun. That is the condition under which behaviour of consumers in the past prompted by dances of that form led systematically to survival and reproduction. There being nectar 400 metres away in the direction of the sun is part of a direct explanation of why behaviour of that type has been stabilized by natural selection. (Millikan also places considerable weight on there being a systematic relationship between different dances and their corresponding locations, which I discuss further in §5.5.)

Millikan coined the term 'Normal explanation' for this kind of evolutionary explanation of how representation-prompted behaviour of the consumer was selected (Millikan 1984). What is evolutionarily Normal may be statistically rare, for example that a sperm actually fertilizes an ovum. The Normal cases are the ones that matter for natural selection. A complete Normal explanation would go into all kinds of details about the mechanism, and might also mention various background factors, like gravity remaining constant. Millikan's focus is the least detailed Normal explanation of the specific type of behaviour prompted by a representation R. For the bee dance, this mentions the presence of nectar 400 metres from the hive, but not details of the implementational mechanism, nor the fact that gravity remained constant.

Consumer behaviours will generally have a nested set of evolutionary functions: to fly off a certain distance and direction, to forage there, to obtain nectar, and to promote survival of the hive and production of offspring. Not all of these figure in the content-fixing story. It follows from what Millikan says about consumers making use of mapping rules that there will be different Normal explanations of different behaviours prompted by different representations. So, we can't simply explain all the dances by the fact that there was nectar somewhere nearby when they were selected. Content is fixed relative to the behaviour of the consumer specific to each representational vehicle. That excludes general evolutionary functions of a behaviour like promoting survival of the hive. For the same reasons there is considerable specificity in the success condition associated with each type of behaviour: getting nectar 400 metres away rather than just getting nectar.

In short, teleosemantics founds content on evolutionary functions, and standard teleosemantics also depends upon there being a special kind of causal structure, a separation between representations and their consumers. Teleosemantics is the basis for a good account of content in some simple representational systems,[18] of which animal signalling cases are a paradigm: macaque alarm calls, beaver tail slaps for danger, and the honeybee nectar dance.[19]

1.5 Challenges to Teleosemantics

Teleosemantics may be a good theory of content for animal signalling, and perhaps also for some kinds of internal communication that are the direct products of natural

[18] Even there, in my view standard teleosemantics needs to be supplemented with a further requirement, so that it is not just an output-oriented theory of content (Shea 2007b). The requirement is that a representation should carry correlational information about the condition it represents (more carefully: that the putatively representational state should have carried correlational information at the time selection was taking place).

[19] Ethological work on animal signalling identifies exactly the same factors as relevant to the content of an animal signal: what the signal correlates with, the behaviour that is produced in response, the evolutionary function of that behaviour, and the conditions that matter for fulfilling that function (Searcy and Nowicki 2005, p. 3).

selection, like hormonal signalling,[20] but there are significant obstacles to applying the theory more widely, to mental representations in general. The purpose of this section is not to demonstrate that teleosemantics fails but, as with the other theories I have mentioned, to say what its main challenges are thought to be, so that the accounts of content I put forward in later chapters can be assessed against those challenges.

The first obstacle to consumer-based teleosemantics is the need to identify a representation consumer with the right properties to play a content-constituting role (both to make it the case that some internal states are representations and to fix their contents). In real psychological cases behaviour depends on the interaction of several different representational vehicles. There need be no identifiable subsystem that takes this collection of representations as input and produces behaviour as output. Instead, the whole organism relies on interactions between internal representations in order to initiate and guide appropriate behaviour. We could instead take the outputs to be internal: representations rather than behaviour. Then each layer of processing could act as a consumer for representations lower down in the hierarchy. But it is not clear whether there is a non-circular account of evolutionary success conditions if the outputs that constitute content are further representations.

Nor does psychological processing always divide into neat layers to allow it to be parcelled up in this way (as we will see in Chapter 4). Some of the most compelling support for realism about representations (for RTM) comes from cases where something is known about the neural states that drive behaviour. Representation in the brain is however particularly unsuited to the consumer-based treatment, for the reason we have just seen: it is very hard to see a principled way to identify representation consumers in the brain, if consumers are devices whose output fixes content (Cao 2012). Even idealized diagrams of neural circuitry are notoriously interactive, consisting of a complex mix of feedforward, feedback, and lateral connections, some proceeding in a series of levels or layers, others cross-cutting or bypassing those layers (see Figure 1.5). That is reflected in information-processing accounts of how representations interact to drive behaviour. I don't take this to be a knock-down argument against consumer-based views, but it will be an advantage of my account that I do not have to appeal to consumers to play a content-constituting role (Chapters 3–5).

The second challenge for teleosemantic theories of content is to formulate a notion of etiological function that is suited to playing a content-constituting role. Both Millikan and Papineau appeal to biological functions. Biological functions are based on evolution by natural selection. Most representation-types we are looking at are learnt. They have not evolved directly by natural selection. Millikan argues that new vehicles produced by learning have derived functions, functions that derive from the purpose of the

[20] Also genetic information: it shows that genes carry semantic information, throws light on what genetic information can be called on to explain, and also applies to other forms of inheritance systems, i.e. of signalling between generations (Shea 2007c, 2009, 2011b, 2012a, 2012b, 2013a, Shea et al. 2011).

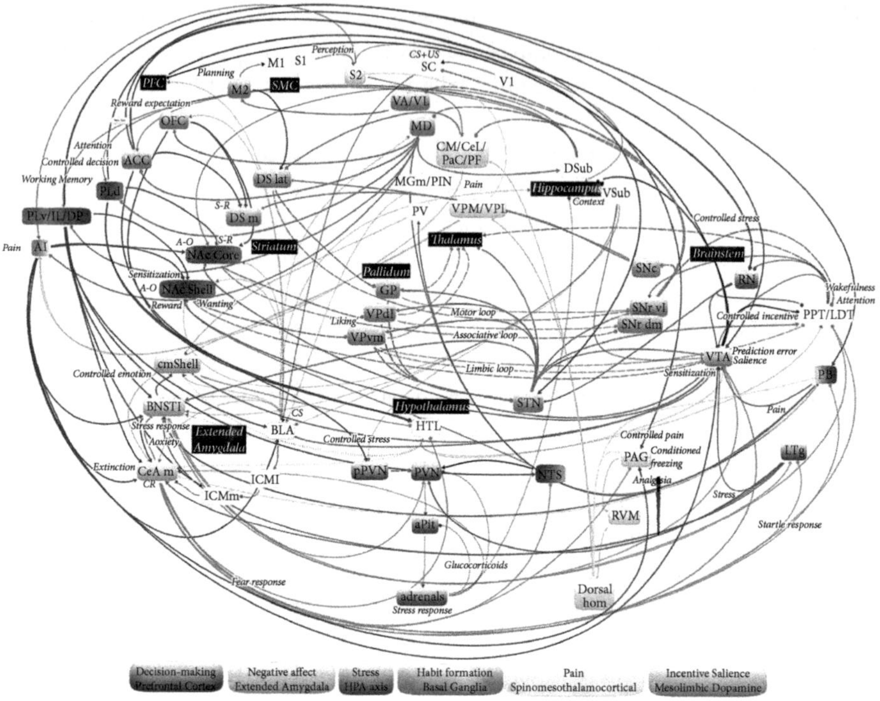

Figure 1.5 Some of the functional connections in the rat brain that are important for reward-guided behaviour (from George and Koob 2010).

learning mechanism. For example, the mechanism in infants that tracks faces and learns patterns in visual input from faces plausibly has the evolutionary function of enabling the infant to reidentify individuals (more precisely, to behave in ways that depend on reidentifying the same individual again, e.g. its mother). That evolutionary function is relational. Which particular individuals it is supposed to track depends on who the baby has been interacting with. So, when the mechanism operates in baby Zeb and learns the pattern that is distinctive of his father Abe, the new representation has the derived function of tracking that particular individual, Abe.

This account works when the learning mechanism has a specific (relational) evolutionary function. But much learning in higher animals, especially in humans, is the result of general-purpose learning mechanisms. For example, the function of classical conditioning is just to find patterns in the sensory input. Such general evolutionary functions do not ground specific functions for acquired representations. Suppose we hear a loon's distinctive song. General-purpose learning mechanisms in the brain can track regularities in acoustic input. So, we learn the characteristic acoustic pattern of the song. On hearing part of the song, the mechanism can complete the pattern. Is this supposed to track loons in general, or an individual loon, or a distinctive pattern in the incoming sound waves or auditory neural input? The very general evolutionary function of the learning mechanism does not decide between these. So relational evolutionary proper functions are not a promising basis for grounding all content.

Millikan argues that functions can also arise directly from learning when this involves a selection process in its own right (Millikan 1984, p. 45, see also Papineau 1987, pp. 65–7). She thinks this applies to instrumental conditioning. Dretske (1988) also puts forward a theory of content based on instrumental conditioning. An animal acts on a stimulus in a new way, which generates a reward, and so cements the disposition to act in that way. The reward is delivered because of some aspect of the stimulus; for example, the light indicated that there would be a peanut on the right-hand side, and the animal learnt to reach to the right-hand side in response to the light. The aspect of the correlational information carried by the stimulus which explains why this action tendency was stabilized constitutes the content of the new representation formed in the process.

Dretske's account is not based on the evolutionary function of instrumental conditioning. It is a basis of content determination, in a system that undergoes learning, that is not derivative from evolutionary functions at all. Nor does it depend on assimilating learning to a process of generate-and-test like natural selection (Kingsbury 2008). It suggests that we should have a broader conception of the kinds of stabilizing processes that can give rise to content. Indeed, Dretske's theory is part of the inspiration for my approach whereby there are several different stabilizing processes that can ground content (Chapter 3). However, Dretske's account of how content explains behaviour only applies to one kind of learning mechanism, instrumental conditioning (Dretske 1988, pp. 92–5; 1991, pp. 206–7). The question for teleosemantic theories of content is to specify which kinds of stabilizing processes give rise to the kind of etiological functions that ground content—and to explain why a certain way of delimiting etiological function is the right one for theories of representation to rely on.

The third main challenge faced by teleosemantics is highlighted by the swampman thought experiment. Swampman is an imaginary molecule-for-molecule duplicate of a person (created, let's say, by random chance when lightning hits a swamp). Teleosemantics implies that swampman has no representational states, because he has no evolutionary history. Some have taken intuitions about swampman to be a basis for objecting to a theory of content. As we will see shortly, intuitions have little probative value for our kind of project (§2.2). Nevertheless, the swampman case is important, because it highlights an implication of the theory. It forces us to reflect on whether there are good reasons for representational content to be based on history.

At first pass, representational explanation does not seem to depend on evolutionary history at all. By recognizing that behaviour was driven by a representation of the location of an object, say, it seems that the cognitive scientist is picking out a synchronic property of the organism. It also seems that the representational vehicle, for example a syntactic or neural state, is a synchronic cause of the behaviour. How internal processing unfolds, and hence how the organism will make bodily movements, is caused moment-to-moment by the intrinsic properties of representational vehicles. It follows that, if we take an organism that has evolved by natural selection and had a lifetime of interacting with its environment, make a duplicate with all the same internal properties, and place the duplicate in just the same environment, we will be able to make the same predictions about how it will behave.

Millikan argues that the intrinsic duplicate falls outside the real kind which underpins our inductive practices (Millikan 1996). Our inductions about people and their representational properties go right, to the extent that they do, because humans form a historical kind, sharing descent from a common ancestor, shaped by natural selection. That answer just pushes back the question however. It doesn't say why there are not also non-historical kinds that would enter into the same kinds of explanation. The fact that predictions would go right for swampman suggests that there is some synchronic property shared by humans, that would also be shared by their intrinsic ahistorical duplicates, and which underpins inductions.

The teleosemanticist should pause at this point and ask us to focus on the explanandum, the thing which representational contents are called on to explain. We point to representations to explain how organisms and other systems manage to interact with their environment in useful and intelligent ways. The explanandum is a pattern of successful behaviour of a system in its environment. That explanandum is absent at the moment swampman is created. It's not just that swampman has not yet performed any behaviour. (He already has dispositions to behave in certain ways.) It is that it's quite unclear that some behaviours should count as successful and others not. So, the creature with no history has no contents but that is fine because it has nothing which contents are called on to explain.

The 'no explanandum' argument does not rescue standard teleosemantics, however (see §6.4). It may show that we have no explanandum at the moment of swampman's creation, but it does not show that deep evolutionary history is needed for there to be an explanandum in place. As soon as an intrinsic duplicate starts interacting with its environment, stabilizing processes will begin to operate. It will do things that contribute to its persistence as an organism. It will undergo learning: behavioural patterns will be repeated or altered based on feedback. Doing things again that have been stabilized in the past looks like a kind of success—these are behaviours that have contributed to survival of the organism and the persistence of these behavioural dispositions (in the recent past). So, an organism's individual history seems to be enough to set up an explanandum which representational contents could be called on to explain.

Dretske's learning-based theory of content calls for individual learning history but not evolutionary history (Dretske 1988). It shows that there being something to explain—such as how the organism succeeds in getting food—does not depend on evolutionary history. So, an organism's learning history, taken on its own, seems to be enough to ground: (i) an explanandum, concerning the organism's interactions with its environment; and (ii) a kind of representational content suited to explaining those interactions. What thinking about swampman shows us is that teleosemantics lacks a good account of why—given its explanatory role—representational content should be the kind of property that depends on *evolutionary* history.

Finally, I come to an objection that can be made to some teleosemantic accounts of content, and which also applies to varying extents to other naturalistic theories of content. How does content get its explanatory purchase? What does it add to a purely causal description of how a system operates and how it interacts with its environment to label some of its states with content? Dretske gave an answer to this question, arguing that contents figure in 'structuring cause' explanations, explaining why a system is wired the way it is, rather than synchronic causal explanations (Dretske 1988). That is an exception, however. Most theories of content, while telling us how content is determined, have relatively little to say about why content determined in that way has a special explanatory role (e.g. Fodor 1991). We turn to this issue in the next chapter, which sets out a framework for content-determination specifically designed to elucidate the explanatory role of content. We return to it again in Chapter 8 once we have detailed accounts of content in hand.

2

Framework

2.1 Setting Aside Some Harder Cases

This chapter sets out the framework within which I will develop my account of content (Part II). The focus is on exposition rather than a detailed defence of the approach. Nor do I mount arguments against other approaches. The framework needs to be judged by its fruits—by whether the accounts of content it underpins are satisfactory. Once that has been laid out in Part II of the book, I will return in Part III to discuss others' views and to defend the framework developed here. I start in this section by homing in on the class of representation that forms the target of our enquiry.

Existing treatments of content have mostly started with everyday examples like occurrent beliefs and desires (doxastic states) and other conscious states. These are indeed paradigmatic cases, but they are not the only place where intentionality is mysterious and the need for a theory of content is pressing. In information-processing explanations in cognitive neuroscience the 'information' is in fact representational content (e.g. Franklin and Wolpert 2011, Yang and Shadlen 2007). Indeed, the cognitive sciences in general extend well beyond doxastic states and conscious states. They are shot through with representations throughout. It is perfectly legitimate for these sciences to take the existence of representations—content-bearing physical particulars—for granted. The sciences of the mind have been remarkably successful in predicting and explaining both behaviour and the goings on in the brain. That success goes a long way towards vindicating the foundational assumption that neural processes trade in mental representations. Nevertheless, the nature of representational content remains puzzling even in non-doxastic, non-conscious cases. It would be a considerable achievement to understand what makes it the case that these states have the kind of intentionality or aboutness presupposed by scientific theories.

To do that I will set aside some features of everyday representations that make the content question more complex. Consciousness is one. I won't deal with cases where a representation's being conscious is relevant to fixing its content. A second feature of ordinary occurrent beliefs and desires[1] is that they enter into epistemic relations: perceptual states justifying beliefs, beliefs justifying other beliefs, and so on. I set aside cases where a representation's entering into relations of justification for the person are relevant to its content. A third feature, which is probably related, is that they are offered as reasons in the social process of explaining to others what we believe and why we act as we do. When acting together, these reasons or verbal reports feature heavily in deliberative joint decision-making. A fourth feature is the kind of constituent structure characteristic of natural language; for example, if there were a kind of representation only available to those capable of using genuinely singular terms.

To have a suitable shorthand, I will use the term 'subpersonal' to cover representations for which content-determination does not depend on those complicating features: consciousness, justification for the person, a role in reason-giving interactions with other people, or being structured like natural language sentences. I am not concerned with the question of whether there is a fundamental personal–subpersonal distinction; for example, a division between incommensurate schemes of explanation (Hornsby 1997, 2000). Nor is 'subpersonal' supposed to connote a distinction between the whole organism and its parts. I use the term simply as an umbrella term for disclaiming these four complicating factors.

Considerable progress was made on the content question in the 1980s and 1990s. In the time since then we have learnt much more about the way mental representations are realized in neural processes. A standard line then was that we would never discern vehicles of content amongst the messy workings of the brain (Fodor 1974, 1987a). The representational level of explanation was taken to be autonomous from the neural to the extent that we should theorize about it separately, without looking to facts about neural realization as a substantial evidential constraint. More recent progress in uncovering the neural basis of representation gives the philosopher of content some fantastic material to work with: cases where a well-confirmed account of the computational processing that generates behaviour can be married with a detailed understanding of the processes in which neural representations are realized. Two cases where there is a convincing computational explanation realized in a well-supported neural mechanism are the neural mechanisms for accumulating probabilistic information about reward (Yang and Shadlen 2007) and the neural circuit responsible for motor control (Wolpert et al. 1998, Franklin and Wolpert 2011). Figure 2.1 illustrates the latter case. The details need not detain us; it is enough to notice the characteristic pattern of explanation: the circuit is described both neuro-anatomically, and also computationally—in terms of the representational contents carried by those neural areas and the way their interactions

[1] Standing beliefs and desires are not conscious, but it is not clear whether there are representational vehicles for these contents.

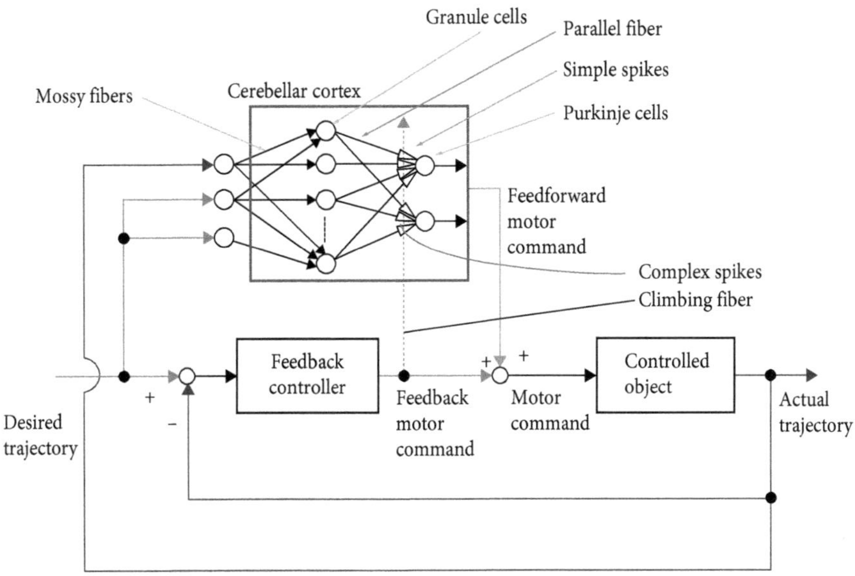

Figure 2.1 Diagram of a leading theory of motor control (Wolpert et al. 1998), used here to illustrate what is characteristic of neural representations. Prescinding from the details, neural areas are picked out both anatomically and in terms of what is represented and computed.

compute a useful function. Because of these developments, we now have a wealth of empirical data against which to formulate and test theories of neural representation.

Since non-conscious, non-doxastic neural representations raise the problem of content in a clear way, one central aim of the book is to provide a theory of content for them. Neural representations form the subject matter of some of our central case studies. Although I'm keen to endorse the idea of neural representation and show that it makes sense, calling the book 'Representation in the Brain' would be misleadingly narrow. The same issues arise in other parts of cognitive science, where there are good reasons think representations are causally interacting physical particulars, but where their neural realization is unknown and could prove intractable. Many of these cases are also subpersonal in the sense of lacking our complicating features. So 'Representation for Cognitive Science' is a better description. Cognitive science does of course also encompass doxastic and conscious states, so my account is not intended to apply to all of cognitive science. What I'm aiming for is an account that applies widely within the cognitive sciences, and which underscores the legitimacy of its reliance on a notion of representation. Hence: *Representation in Cognitive Science.*

My overall philosophical strategy, then, is to start with the subpersonal and work upwards. This inverts the normal approach.[2] But the normal approach has not been entirely successful to date. If we are puzzled about how there could be space in the

[2] Cf. Karen Neander's recent book which also focuses on simpler cases. Her target is non-conceptual representations (Neander 2017, pp. 27–46).

natural world for intentionality at all, then seeing how it arises in a range of cases in cognitive science will be a major step towards resolving the puzzle. Furthermore, seeing how representational content arises and earns its explanatory keep in these cases should prove a useful staging post on the way to tackling the more complex cases. So, an account of subpersonal representational content is part of a broader strategy for tackling the problem of intentionality. Given the starring role of representational notions in the cognitive sciences, it would also be a significant result in its own right.

2.2 What Should Constrain Our Theorizing?

The most widespread way to test theories of content has been to line them up against intuitions about what mental states represent. That method reached an impasse of conflicting and often theory-driven intuitions. Especially when focusing on the subpersonal, it is clear that intuitions should be given little weight. For personal-level mental states, like beliefs and desires, we have some reason to rely on our intuitions about content, our judgements about what our thoughts mean. Even there, experimental results on the unreliability of intuition when people are giving common-sense explanations of their behaviour should be make us cautious (Nisbett and Wilson 1977, Johansson et al. 2005, Carruthers 2011). When it comes to subpersonal representations, it is unclear why intuitions about their content should be reliable at all.

I take a different approach. A theory of content is answerable not to intuition, but to the role representations play in explaining behaviour. The rat finds its way back to food in a maze because it accurately represents its position and the location of food. Representing correctly explains successful behaviour and misrepresentation explains failure. A good theory of content should show how the contents it specifies are suited to explaining behaviour in that way.[3]

To do that we need to examine a range of cases where subpersonal representations explain an organism's outputs.[4] Experimental psychology and cognitive neuroscience give us a large number to choose from. We'll mostly look at the behaviour of organisms, but artefacts like computers and control devices also produce outputs in response to their environments as a result of representation processing. I'll use 'behaviour' as a neutral term for actions and other outputs (but not internal workings) and 'system' as an umbrella term for organisms and other entities whose behaviour is representationally generated.[5]

When a scientific explanation points to representational content to explain behaviour, we need to get inside that explanation to see how it works. That means getting into the

[3] I remain neutral on whether representation has further explanatory roles; for example, explaining why internal processing unfolds in a certain way.

[4] In all of our case studies, outputs are actions and their consequences; but other kinds of output can also be considered; for instance, physiological, hormonal, and neurochemical outputs.

[5] This includes some systems that have organisms as subsystems. The honeybee colony is a system in this sense. An individual consumer bee is a subsystem of this system.

details of a behavioural profile and its causal underpinnings. Only with the details in view can we properly address the question: what kind of thing is representational content, to enable behaviour to be explained in that way? We should be open to finding cases where the attributed representational content is doing no work—it is just a way of understanding the system, a useful fiction—or where content is different from that attributed by the psychological theory. If our conclusions are going to have general applicability, we need to pursue a wide range of case studies so as to sample systems that have different features and operate in different ways: perceptual, motoric, and cognitive, in humans and in other animals. So, a seemingly innocuous desideratum—that representational content should be characterized by reference to its explanatory role—turns out to be a prescription for getting into the details of a wide range of case studies from subpersonal empirical psychology. That is exactly what we will do in this book.

The project is to put forward one or more theories of content that tell us how the representations involved in these case studies get their contents. Rather than intuition, our theorizing is constrained by a desideratum, something we want to explain. What we want to explain is how adverting to representational content allows us to explain behaviour. We want an account of content which shows why content plays that proprietary explanatory role:[6]

> *Desideratum*
> An account of how representational content is constituted in a class of systems should allow us to show why recognizing the representational properties of such systems enables better explanations of behaviour than would be available otherwise.

Since we are investigating cases where there are real vehicles of representational content, individuable non-representationally, a causal account of the operation of the system in terms of vehicle properties will always be available in principle. The vehicles in a computer are electrical currents in semi-conducting chips, which causally interact in virtue of their electrical properties; similarly for neural vehicles consisting of patterns of neural activity. Neural activity unfolds in virtue of electrical and chemical properties of neurons and synapses. Vehicle properties depend only on intrinsic physical properties of the system, its parts, and their interrelations. So, any interaction with the distal environment could in principle be 'factorized' into three components: the way the environment causes changes to intrinsic physical properties of inputs to the system; the way those inputs cause changes to other internal states of the system, issuing eventually in motor movements produced by the system; and the way those movements of the system cause changes to its distal environment. To meet the desideratum,

[6] I call this a desideratum, rather than a necessary condition on the existence of content. If it were not met, it's not clear that we would be forced to give up on there being representational content, rather than changing our expectations about the nature of content.

representational content has to offer a better explanation of behaviour than such a 'factorized' explanation can provide (see §8.2).

To take an example from Ramsey (2007, pp. 138, 140–1), a rifle responds to a finger movement by discharging a bullet from the muzzle. There is an internal mechanism whereby movement of the trigger (input) causes movement of the firing pin, causing ignition of the primer in the cartridge, causing explosion of the propellant, causing the bullet to travel down the barrel and exit the muzzle at speed (Figure 2.2). The movement of the firing pin is designed to correlate with movement of the trigger finger at input, and to lead to the firing of a bullet at output. A teleosemantic theory, based on deliberate design rather than evolutionary function, could treat the cartridge as a 'consumer' for the state of the firing pin, implying that when the pin moves it represents *the trigger has been pressed, fire a bullet.* However here a representational explanation of the behaviour of the rifle would march exactly in step with a factorized explanation that simply describes the causal chain from finger to trigger to firing pin to primer to propellant to bullet—without mentioning content at all. The widely used example of magnetotactic bacteria is also a case where a factorized explanation marches exactly in step with the putative representational explanation. As standardly described in the philosophical literature the bacterium contains a little magnet which causes the whole organism to align with the direction of the earth's magnetic field hence to swim in that direction (Dretske 1986): magnetic field causes alignment causes direction of motion. We will see that the bacterium will not satisfy our conditions for having representational content (§8.2b).

Our desideratum is to show why representational content allows for better explanations than would be available otherwise. Theorists often make a stronger demand: that a theory of content should show why representational explanations are indispensable,

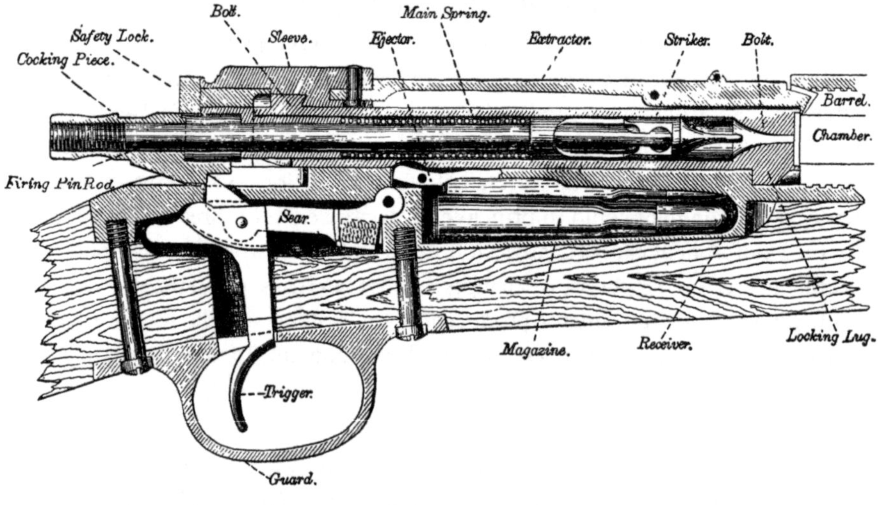

Figure 2.2 The mechanism for firing a rifle (an example discussed by Ramsey 2007).

allowing us to explain something that could not be explained otherwise (Dennett 1971). That is too strong. If appealing to representations offers a better or more perspicuous explanation of behaviour than would be available otherwise, that is a sufficient motivation for adverting to representations when explaining behaviour. This chapter puts forward a framework designed to meet the desideratum, but it is not until we have the positive accounts of content in hand that I will be able to show how content-based explanation captures something important that the factorized explanation misses (§3.6 and §8.2b).

I am not aiming to analyse the concept of representation: the everyday notion or the scientific notion. The theories discussed in Chapter 1 are often rejected as failing to accord with intuition. My test is whether they meet our desideratum. I take my task to be to define some technical terms and show that they are useful. Various kinds of content are defined in the chapters to come (UE information and UE structural correspondence, both based on a variety of task functions). What I aim to show is that these terms are non-empty and that the properties they pick out are useful for explaining behaviour. They are properties of internal vehicles that allow for the explanation of successful and unsuccessful behaviour of an organism in terms of correct and incorrect representation. This does not imply that an organism only has content when it would be useful to treat it as having content. That is a different kind of theory, one which makes the very existence of content relative to its explanatory merits for an interpreter. I do aim to show how content, as I define it, gets a useful explanatory role in general, but that does not imply that every instance of content is explanatorily useful, nor does it make the existence of content relative to an interpreter. The properties defined by the terms I will introduce exist whether or not anyone is around to make explanatory use of them.

2.3 Externalist Explanandum, Externalist Explanans

As we saw in the last chapter, underlying the idea that the mind processes mental representations (RTM) is a core insight: mental representations are physical particulars which interact causally in virtue of non-semantic properties (e.g. their physical form) in ways that are faithful to their semantic properties.[7] Psychological processes like thinking, perceiving, planning, reasoning and imagining consist of causal processes taking place between representations with appropriate contents. Here I explain how I think that insight is best understood.

RTM is committed to there being real, individuable vehicles of content. The problem we have just seen then becomes pressing. A complete causal account of the operation of the system will be available in non-contentful terms. Proximal stimulation at input will cause the system to undergo various internal transitions that eventuate in movements at output. Intermediate entities at various stages of this process may *have* semantic

[7] §8.3 discusses the causal efficacy of semantic properties.

properties, but content does not figure in the underlying causal story about internal transitions and bodily movements. We can see this very clearly when we look at a computational system designed to calculate how money flows through the UK economy. Moniac is a computer that uses water in tanks to represent money. It is obvious that the various levels of water representing national income, imports, tax, and so on interact solely in virtue of the physical properties of water and of the mechanisms it flows through (Figure 2.3). The vehicles are much harder to observe in psychological cases, but the principle is the same—a non-semantic causal story is always available.

Contents come into view when we target a different explanandum. An organism interacts with the environment, bringing about distal effects in its environment, doing so by reacting to distal objects and properties in the environment. There are real patterns in the environment and an agent's interactions with it that would be invisible if we looked only at intrinsic properties of the agent.[8] Those patterns call for explanation. They are an additional explanandum, beyond the question of how processes within an organism unfold over time—an explanandum concerning the organism's interactions with its environment. Given an externalist explanandum, externalist properties of the system and its components are good candidates to figure in the explanation (Peacocke 1993). Which externalist properties? Plausibly content properties, if contents are externalist. Representational contents figure in explanations of how an organism interacts with its environment and achieves distal effects in its environment. So it makes sense that content should be externalist, determined partly by extrinsic properties of representational vehicles (cp. Ramsey 2007, pp. 95–6). Contents would then be suited to explaining real patterns that are the result of organism–environment interactions. They would then explain things which purely intrinsic properties of the system do not.

Not every possible interaction between a system and the environments it could be placed in calls for explanation in this way. The way a spider is swept down a river does not. Nor does every system enter into the sorts of interaction that representations are characteristically called upon to explain (none of a river's interactions call for the river to be a representer). I take from teleosemantics the idea that it is when a system is performing a function (e.g. getting nectar for the hive) that representational explanation becomes pertinent. Or at least, that explaining the performance of functions is one central way that representation gets its explanatory purchase. Chapter 3 develops a more general account, going beyond evolution by natural selection, of which input–output behaviour generated by an organism counts as functional. I call these 'task functions'. The important point for us now is that an organism with task functions achieves distal effects in its environment, can do so successfully or unsuccessfully,

[8] I adopt Dennett's catchy terminology without aiming to capture exactly what he meant by 'real patterns' (Dennett 1991). For me, real patterns are observer-independent approximate regularities occurring at some level of description whose existence allows us to describe the system in a more compact way at a more coarse-grained, less fundamental level (cp. Ladyman and Ross 2007, Ladyman 2017). Being more fundamental is a matter of applying at more scales (of length, time, or energy).

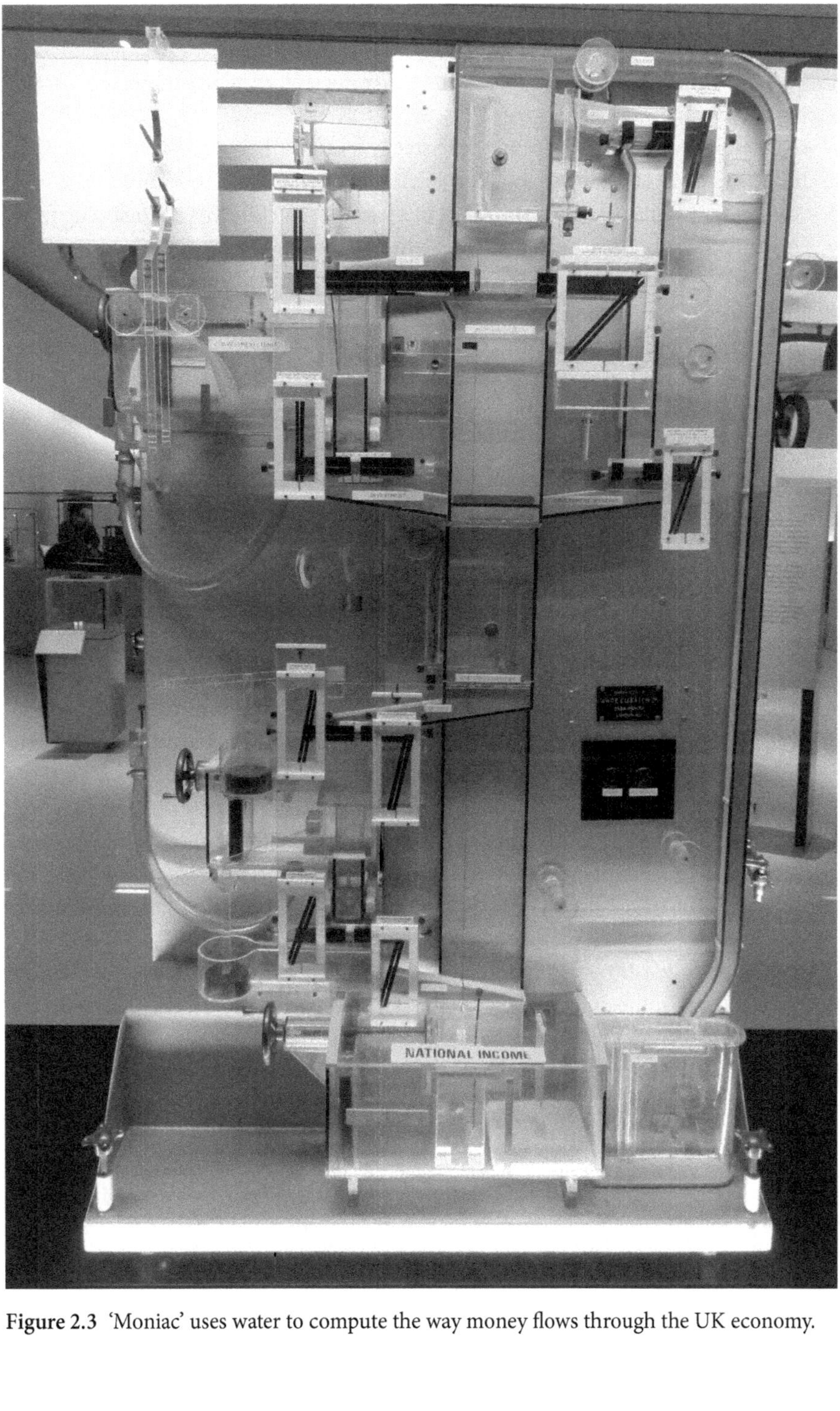

Figure 2.3 'Moniac' uses water to compute the way money flows through the UK economy.

and does so by reacting to and interacting with distal objects and properties in its environment. A task function performed by an organism comprises an explanandum, about an organism's interactions with its environment, to which representational explanations can be addressed.

A task function is mapping from some states to others. For example, a rat can take itself from a range of locations to a new location where there is food and ingest it. A person can take some materials and make them into a coat. The first is a mapping from one organism-environment relation to another, the second a mapping from one state of the environment to another. Both are mediated by the activity of the organism. For a given function, there are many possible ways to perform it. Doing arithmetic, people perform a function that takes written numbers as input (say) and produces a written number, their product, as output. There are many different ways of achieving that input–output mapping. I might use long multiplication in columns with carrying. That is a way of producing the appropriate input–output mapping. The same is true for moving from one location to another or transforming some materials into a coat. Those mappings between states can be achieved in multiple ways, and there is a fact of the matter about how a particular organism does it.

David Marr's algorithmic level specifies the way a function is carried out by a system (Marr 1982). Long multiplication in columns is an algorithm for multiplying together any two numbers. Stretching the term slightly, I am going to use the term 'algorithm' for the way an organism carries out functions of the kind I have just described (navigating its environment, getting food, making tools, etc.). In the multiplication case the organism is calculating the input–output function (from two numbers to their product), but in the other cases the idea is looser. The organism carries out an algorithmic process over representations in a way that leads to it achieving the function (e.g. getting from one location to another and ingesting the food there). In my sense an algorithm is a sequence of operations between representations that leads to an organism performing a function.[9] The sequence of operations is a computation carried out by the organism.[10]

In my usage algorithms are concrete. They are a way of processing representations that is realized in some organism or other system. An algorithm described in terms of transitions between contents is neutral about how those transitions are realized, save that there must be states of the system that carry the appropriate contents and undergo the

[9] The stipulation that algorithms must involve representations is just for convenience of exposition. I am not assuming a semantic view of computation individuation. A sequence of operations over non-semantic states that can be carried out in a finite amount of space and time could also be described as algorithmic. On some views it could count as a computation irrespective of whether anything is represented. In that sense the rules for the flow of activity in a connectionist network would count as algorithmic, as would the learning rule. Which kinds of connectionist processing count as algorithmic in my sense depends on how it is appropriate to describe them representationally (Shea 2007a).

[10] Some theorists reserve the term 'computation' for processes that deal in discrete states (Eliasmith 2010, p. 314), whereas others use it more broadly, so that the idea of analogue computation is not self-contradictory. I adopt the broader usage, covering all cases where representations are physical particulars, processed in virtue of their vehicle properties in ways that respect their semantics.

appropriate transitions. Those transitions must also be describable non-semantically, in terms of the way the system moves between physical states—what is often called a syntactic description. It is this constraint which makes for a realist answer to the question: which algorithm is system S using to perform function F? Steps of the algorithm have to map onto causal transitions in the internal processing going on within the system.[11] There are several ways the problem of achieving a given input–output mapping could be broken down into intermediate steps (see Figure 2.4). A given system uses one of them. Vehicle realism means that the representational explanation of behaviour is an account of the particular way a system achieves a given input–output mapping, hence of how it manages to perform a task function.

In most examples from the cognitive sciences the inputs that the system responds to and the outputs being explained are not intrinsic features of the organism, but are partly externalist.[12] Suppose we have a system that has been trained to track the direction of motion of surfaces and make a movement in the corresponding direction. One algorithm for doing that keeps track separately of the colour and motion of small portions of surface, and then combines that information to infer which portions of surface are parts of the same surface and what the overall direction of motion of each surface is (§4.7). Steps of the algorithm are described in terms of their representational content, such as representing the colour of the surface-portion at such-and-such location. Processing a series of representations with those contents is the way the system produces the distal input-output mapping.

As we just saw, if contents are going to explain how a system performs a distal function, we should expect content to be determined in part by extrinsic properties of the vehicles of content: relations those vehicles bear to objects and properties outside the system. Which relations? Here I draw on Peter Godfrey-Smith's idea that representations bear exploitable relations to features of the environment (Godfrey-Smith 2006). Godfrey-Smith takes that to be part of his 'basic representationalist model', but the

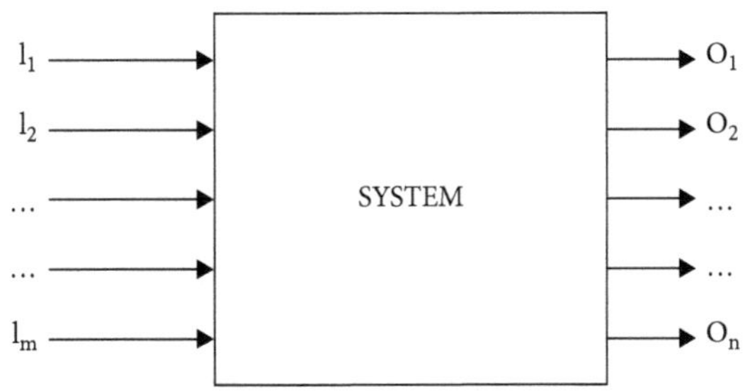

Figure 2.4 The input–output mapping produced by a system does not fix what goes on inside the box. The mapping could be generated by a look-up table and usually also by several other algorithms.

[11] The processing has to undergo transitions that are called for by the algorithm, hence appropriate to the contents represented, but that does not imply that the causal processing is sensitive to content.

[12] That feature is not widely noted in cognitive science. It is debated in philosophy (Egan 1991, Segal 1991).

idea is still applicable when we drop the requirement for a representation consumer (§1.5). The entire system processes a variety of representations, possibly in complex ways, so as to carry out a distal task function from some states to others. In order to perform that function, the system makes use of the fact that intermediate components—vehicles of content—bear exploitable relations to distal features of the environment. One useful relation is to have a component that correlates with a relevant feature of the environment (Chapter 4); for example, that correlates with the colour of part of an object. Another useful relation is to have a structured set of components whose structure mirrors an important structure in the environment (Chapter 5); for example, to have a cognitive map of the spatial environment. The system as a whole makes use of these exploitable relations in calculating how to behave.

To be an implementation of an algorithm for performing a system's distal functions, internal components must have two kinds of properties at once. Causal transitions between vehicles must be those called for by the algorithm. That is a matter of the intrinsic properties that drive internal processing. And components must also have extrinsic properties that give rise to the contents which are called for by the algorithm. (How that can be so is the subject of the rest of the book.) Those contents must be respected when the vehicles are processed: intrinsic properties of the vehicles and the system in which they are processed must be such that the transitions between vehicles make sense in the light of the vehicles' relevant extrinsic properties. Exploitable relations are the link between internal components and the distally characterized task function which the organism is performing. It is the coming together of extrinsic properties and intrinsic properties in this way that gives rise to content.[13]

2.4 Representation Without a Homunculus

One alluring but mistaken way of thinking about mental representations is as being inner sentences that are understood by some internal homunculus. The model is the way we understand external sentences: we hear the words and attach meaning to them. It is a mistake to think we do something similar with mental representations: that when they occur in the mind we have to look up their meaning before we can reason or act appropriately. That would require some kind of inner interpreter of the mental representation, launching a regress.

The 'homuncular functionalist strategy' avoids the regress (Dennett 1978). First off let's see how this works if we presuppose representation consumers, before generalizing the insight to my account. The consumer of a representation does not understand its meaning. It is just disposed to react to a representation by producing a certain behaviour. The consumer is not acting that way because the representation has a certain meaning. Rather, the fact that the consumer behaves in that way constitutes the representation as

[13] ...gives rise to the kind of content we are investigating here (see point about pluralism below). That caveat is implicit throughout.

having a certain meaning. Consumer honeybees don't need to understand the dances they observe; they just need a causal disposition to fly off to the corresponding location. Their behaviour constitutes a dance as having a certain meaning.

The homuncular functionalist strategy is to show that a complex mental capacity arises from the interaction of simpler components, bottoming out in components whose causal operation does not presuppose any mentality or intentionality. My account of content (Chapters 3–5) does not rely on content-constituting consumers. But it still makes use of the homuncular functionalist strategy. Content arises out of the fact that a system has a certain kind of internal organization and is performing a certain function. Nothing in the system needs to interpret the internal representations or understand their content. Later chapters contain detailed proposals for the way content arises when functions, internal processing, and exploitable relations come together in the right way. These are all perfectly un-mysterious natural facts about an organism, computer or other system. The system's interactions with its environment have stabilized the performance of certain functions. It has a certain internal organization. Components correlate with or correspond structurally with distal features of the environment. If content is constructed out of these properties, as I claim, then content properties arise automatically when a system of the right kind is embedded in the right kind of task environment. They don't presuppose an internal understander of representational content.[14]

2.5 What Vehicle Realism Buys

In cases where realism about representation is justified there is a further question about whether it offers any explanatory advantages. This section offers a perspective on that question which fits with the account to come.[15]

According to RTM, transitions between mental representations are faithful to their contents, amounting to an algorithm by which a system achieves its functions. It is not just by chance, or through a look-up table, that the system instantiates a useful input–output mapping. It implements an algorithm and, if more than one algorithm would produce the same mapping, there is a fact of the matter about which one the system is in fact using. That is underpinned by the commitment to there being real vehicles of content. The algorithm must operate over a set of vehicles that can be individuated non-semantically and must follow a series of processing steps that can be specified non-semantically. Thus, vehicle realism is needed for representations to be involved in explaining how a system achieves its task functions in the way discussed above (§2.3).

[14] Nor does content depend on an interpreter in a second sense: an external interpreter that treats the system as having contents. Having content of the kind described here depends on having a certain complex of observer-independent properties. Systems that have these properties are susceptible to a special scheme of explanation, but being so-explicable is not what makes it the case that a system has contentful representations: see §4.2b and §8.5a.

[15] I return to these issues in Chapter 8.

That is the first of three explanatory advantages of vehicle realism: phenomena that it allows us to explain in a distinctive way.

The second is that it predicts a pattern in the way errors propagate. Not only does correct representation explain success and misrepresentation explain failure, but we can make predictions about patterns of failure. An incorrect representation will produce consequences in downstream processing: one error will propagate and lead to other errors downstream, errors that make sense in the light of the misrepresented content. Correlatively, parts of processing that occur before the misrepresentation, or are insulated from it, will not be driven into error thereby. Consider a mechanism that computes object motion by first representing the colour and local motion of small portions of a surface, and then integrating that information into representations of moving surfaces. A misrepresentation at an early stage, such as about local colour, is likely to lead to an error about global motion at a later processing stage. The converse is not true (in a case where only feedforward processing is involved): an error introduced at the stage of computing global motion will not cause errors at earlier stages, such as representing local colour. Ascriptionist views about content do not predict these kinds of systematic relations between incorrect representations.[16] If representations were not real entities in the system, individuable non-semantically, we would lack a ready explanation of why errors should propagate and pattern in these ways.

Thirdly, vehicle realism explains a familiar pattern of stability and change in representational capacities over time. A system tends to keep the same representational resources over time and, when they change, representations tend to be gained and lost piecemeal. Exploring an environment, we learn about new locations one at a time. If representational contents were just attributed as a useful summary of behavioural patterns, it would be unclear why changes to the system's behavioural dispositions should go along with piecemeal rather than wholescale changes to the ascribable contents. In cases where this phenomenon is observed empirically, the representational realist has a ready explanation in terms of the gain and loss of representational vehicles.

Those three patterns of explanation depend on realism about representation: upon there being a substantial non-semantic sense in which an individual token counts as being the same representation again.[17] What makes different tokens instances of the same representation is that their non-semantic properties ensure they are processed in the same way by the system. So representational vehicles can be individuated non-semantically, in terms of intrinsic properties of the system.

[16] NB Davidson's and Dennett's views were not intended to apply to subpersonal representations.

[17] Neo-Fregean senses would explain some phenomena of the second and third type (patterns of error and piecemeal change). (Recall that I'm leaving aside contents at the level of sense, if they exist, and just focusing on referential content.) Senses won't replace vehicles. Vehicle realism is still needed to secure the first explanatory advantage. It is also needed to explain differences that go beyond modes of presentation, e.g. between different people who grasp the same sense, or a single thinker failing to identify two representation tokens which have the same sense. I don't here get into the converse question of whether the idea of vehicles and syntactic types allows us to do without senses (Millikan 2000, Sainsbury and Tye 2007, Recanati 2012).

Care is needed with the idea of a vehicle. A vehicle is the particular that bears representational content. Words on the page are vehicles of content. In referring to the word we don't just pick out ink marks on the page, we individuate them as falling under a type. The word 'barn' is a type that includes 'BARN', '*barn*' and 'barn'. However, the way those marks get their meaning depends not just on their intrinsic shape, but on what language they're being used in. The string of letters 'barn' in Swedish means *child* not *barn*. I will use the term 'syntactic type'[18] for the way of typing vehicles that aligns with content assignment: same syntactic type ensures same content.[19] For a case like 'barn' we would commonly say that the same representational vehicle means different things in English and Swedish. So, vehicles are not the same thing as syntactic types. The same vehicle can fall under different syntactic types in different contexts. Syntactic typing depends on the way a vehicle is processed. Analogously, the vehicle 'barn' is processed one way in English and another in Swedish. We are not looking at natural language and in our cases the way a vehicle is processed depends only on intrinsic properties of the organism/system doing the processing. So, although syntactic type need not be an intrinsic property of the representational vehicle, syntactic types can be individuated in terms of intrinsic properties of the system.

In short, vehicles are individual bearers of content picked out in terms of intrinsic processing-relevant non-semantic properties; and syntactic types are ways of typing vehicles into non-semantic types that are processed the same way by the system, and so are guaranteed to have the same content. In the brain, a distributed pattern of firing in a cortical layer can be a vehicle of content. Neural reuse means that the same pattern of firing may be put to different uses and processed differently when the organism is performing different tasks. So, the same neural vehicle (pattern of firing) may fall under different syntactic types as its effective functional connectivity changes. It may represent spatial location when processing is set up one way, and episodes in the organism's past when processing is set up another way.

Recall the dual nature of content (§2.3). Content arises from the convergence between an externally specified function being performed by a system and internal processing which implements an algorithm for performing that function. It follows that whether an internally specified state counts as a vehicle of content at all depends in part on the system's environment. Being a representation is not dependent only on intrinsic properties of the system. Syntactic typing is therefore partly externalist. In Shea (2013b, pp. 504–7) I give an example where what counts as one vehicle for one task divides up into many vehicles when the system is performing a different task. Is syntactic externalism compatible with the explanatory advantages which I have just claimed follow from vehicle realism? Yes, because it still follows that instances of the

[18] Syntactic does not here imply that the representation has to have constituent structure. It connotes that aspect of syntax that is about individuating the content-bearers, doing so non-semantically. Given the problems with the term 'vehicle' there seems no better term than 'syntactic type' for the non-semantically individuated typing to which contents are assigned.

[19] I am leaving aside problems of indexicality (§8.6).

same syntactic type *within a given system* will share processing-relevant intrinsic properties. That is what is needed to secure the advantages of realism: to give reality to the algorithm, to predict relations between errors within a system, and to explain stability and piecemeal change of representational resources in a system over time. Which intrinsic properties count as syntactic types in a given system will however depend on factors extrinsic to the system.[20]

Syntactic types can be based on properties of dynamical processes. Indeed, neural firing rate is a dynamical property. Dynamical systems theory is used to launch many different arguments against the representational theory of mind but, taken alone, the observation that dynamical processes are responsible for generating behaviour does not in itself undermine representationalism. Elements in a dynamical system can have vehicle properties that are calculated over so as to implement an algorithm for producing appropriate behaviour. To take an imaginary example, suppose walking depends on synchronizing the frequency of two dynamic loops, one involving each leg, the brain involved in both. The oscillation frequency of one leg-involving loop is not fixed by properties of motor neurons alone. It also depends on the weight of the leg, the physical properties of the bones and muscles, how they are coupled to each other, and their coupling to the brain via outgoing and incoming nerves. The phase offset between the oscillations in the loops for the right and left legs could be a vehicle of content; for example, an imperative representation of whether to speed up or slow down. It could interact in internal processing with other dynamic vehicles; for example, the recent rate of energy depletion (acting as a representation of urgency). There, dynamical properties would interact in ways that are faithful to representational contents.

It is of course a substantial question whether a dynamical system is a representational system and whether any dynamical properties are the basis for syntactic types. Useful behaviour can be achieved by some appropriately organized dynamical systems without any representations being involved at all. However, our framework applies very readily to dynamical cases and there is nothing in dynamicism as such which counts against dynamical properties being vehicles of content. Dynamical parameters like frequency, resonance, phase, impedance, and gain are all candidates.

I end this discussion of vehicle realism with a brief note about the underlying metaphysics and its relation to reductive and non-reductive physicalism. One way of naturalizing content is by reducing it to something else: finding a property identity. On the reductive view, having the representational content p is identical to having some (possibly complex) non-semantic, non-mental, non-normative property. That would indeed show, in naturalistic terms, how content is determined. A naturalistic theory of content need not, however, be reductive. It is a familiar point that many special sciences

[20] Oron Shagrir makes the same argument about the nature of computation (Shagrir 2001), where similar issues arise. On his view (a version of the sematic view of computation), whether a system is performing a computation depends in part on factors extrinsic to the system (Crane 1990, Bontley 1998, Horowitz 2007).

deal in properties that are not reducible to a more fundamental science. That is likely to be true of representational content as well.

Non-reductive physicalism is compatible with there being exceptions to the generalizations which connect properties in different domains, with there being *ceteris paribus* bridge laws between different schemes of explanation. So, the account which says how other properties determine content properties could admit of exceptions, provided content supervenes globally on physical properties.[21] A sufficient condition for content determination, although it has nomological force (it is a non-accidental generalization), may admit of exceptions where the condition is satisfied but there is no content, exceptions that can only be explained at some other more fundamental level.

Furthermore, it would be adequate to have a series of different content-determining conditions. Each would be a content-determining sufficient condition[22] applicable to certain cases. That would be enough to show how an appropriate constellation of properties from other special sciences gives rise to content. There is no need to find a single set of necessary and sufficient conditions that covers all possible cases. Naturalism is a substantial requirement but it does not demand that we find a property identity.[23]

2.6 Pluralism: Varitel Semantics

So far I have set out a framework for realism about mental representation. The framework has two variable elements: the source of the distal functions performed by a system; and the nature of the relations that elements in the system bear to the environment, which are exploited in order for the system to perform those functions. The case studies to follow will show how these arise in different ways.

Two types of exploitable relation cover all the cases we will consider: carrying correlational information (Chapter 4) and bearing a structural correspondence (Chapter 5). I am not committed to there being a single overarching property, bearing an exploitable relation, which covers both of these cases without over-generalizing, applying to too many other cases. I don't take it to be an objection to the account that it offers more than one sufficient content-determining condition. So, I present the accounts disjunctively. Exploited correlational information appears in one set of sufficient conditions (conditions for 'UE information', Chapter 4), exploited structural correspondence appears in another (conditions for 'UE structural correspondence',

[21] That is, there should be no content difference without a non-semantic, non-mental, non-normative difference somewhere. A case which counts as an exception to a *ceteris paribus* bridge law should be different in some respect to those which fall under the law.

[22] While aiming only at a sufficient condition, we still want to avoid otiose clauses or unnecessary requirements. Every requirement should be a necessary part of the sufficient condition.

[23] A disjunction of such conditions, even if the list were closed, does not automatically amount to a reduction, since arbitrary disjunctions of properties may not be the right kind of thing to figure in a reductive property identity.

Chapter 5). I am not attempting to define a single overarching technical term that covers both cases. If a definition that would cover both cases applies beyond cases of correlation or structural correspondence, then there is a significant risk that the resulting notion would apply to too many cases. Liberality per se is not objectionable, if it is just generality, but liberality is an objection if it robs content of its distinctive explanatory purchase. Therefore, I carry out the project in a way that is open to pluralism: to content being constituted differently in different cases.[24]

Functions are a second source of pluralism within my framework. Different kinds of function can underpin content. I have already suggested the idea that stabilizing processes other than natural selection can underpin a distinction between successful and unsuccessful behaviour (§1.5). Dretske's case of instrumental conditioning is one example (Dretske 1988). The next chapter argues that at least four different processes give rise to teleofunctions: evolution by natural selection, behavioural learning from feedback, contribution to persistence of an individual organism, and deliberate design. We can recognize that several different processes give rise to teleofunctions without being committed to there being a single overarching theory of function that covers all of the cases—without over-generating and hence robbing the category of its explanatory purchase.

I unite three of these teleofunctional processes under the label 'stabilized function' (§3.4), and all four under the label 'task function' (§3.5). This makes it seem as if I do have a single overarching account of function: task function. In fact, the label 'task function' is just a terminological convenience. Since stabilized functions and task functions have disjunctive definitions, they in effect generate a series of different conditions for content. That is a second source of pluralism, giving us 2 (exploitable relations) x 4 (functions) content-determining conditions. Those conditions bear a striking family resemblance to each other, but I am not committed to there being a single ur-theory which encompasses them all without being too liberal (i.e. which still ensures that there is something distinctive about content explanation). These are all properties that deserve the label 'representational content', but the result of pluralism is that I am not offering a single overarching set of necessary and sufficient conditions for content.

A final source of pluralism is the simplifying move at the start of this chapter: setting aside representations at the personal level. I do think we will need a different theory to account for the content of beliefs and desires, and of conscious states; maybe more than one. But I don't need to make an argument to that effect here. For now my claim is that content may be constituted differently at the personal level, so the accounts offered below should not be tested against the contents of personal-level states.

[24] My pluralism was inspired by Godfrey-Smith (2004), although his pluralism about representation in cognitive science is based on variation in what scientists think the most fundamental basis of meaning is when they apply the basic representationalist model. My pluralism has a different motivation.

I will not be making a positive argument for pluralism. The point of being open to pluralism is that it allows me to resist the pressure to find a single overarching necessary and sufficient condition that covers all the cases. We may get one theory of content that gives us a good account of the correctness conditions involved in animal signalling, say, and another one for cognitive maps in the rat hippocampus. There is no need to find a single account that covers both.

When in the past I argued that a theory of content which combines correlational information with teleofunctions is applicable to some simple cases like animal signalling I gave the combination the label 'infotel' semantics (Shea 2007b). The framework developed here, as well as being different in some respects, is also more widely applicable. A variety of exploitable relations are involved: correlational information and structural correspondence. Indeed, it could turn out that other kinds of exploitable relation exist in other cases. A second source of variation is the range of different functions that can underpin content. So 'varitel' semantics seems like a good term, with the 'vari' marking both variations in the exploitable relations and variations in the teleofunctions involved. The resonance with 'varietal' is apposite, registering the fact that my account of content comes in several non-overlapping varieties.[25]

This chapter has set out the framework of varitel semantics and motivated it as an approach to naturalizing content. It has several distinctive features. Pluralism is one, as is the focus on what I have been calling subpersonal contents. My account does not rely on a representation consumer that plays a content-constituting role. Eschewing intuitions about cases is also a move away from the earlier literature. Although looking at the explanatory role of representation is not new, the desideratum set out above is somewhat distinctive. I also offer my own particular take on realism about representations and its explanatory advantages; and on exploitable relations and the dual nature of content. That is the prospectus. We move now to details of the positive accounts (Chapters 3–5).

[25] My neologism sounds in my head closer to 'vary-tel', a near-rhyme of 'fairy-tale' (although I'm hoping my account is not one of those).

PART II

3

Functions for Representation

3.1 Introduction

Varitel semantics has two variable elements: functions and exploitable relations. Chapters 4 and 5 look at exploitable relations. This chapter deals with functions. To apply our framework, we need to specify what it is for there to be a task being performed by an organism or other system. These tasks are the functions it is supposed (in some sense) to perform. Philosophical work on function has mostly focused on naturalizing biological functions, for which the constraints may be different. We are after a notion of function that is suited to figure in a theory of content: function-for-representation. Philosophical theories of function are often tested against intuitions about what counts as a function, a malfunction, a side effect, or an outcome with no kind of function at all. Intuitions can bear little weight for our purposes. Instead, our theorizing is guided by the goal of explaining representational explanations of behaviour (the desideratum in §2.2).

I take from teleosemantics the idea that natural selection is a source of functions that are partly constitutive of content. However, evolutionary function is too narrow

(§1.5). When behavioural dispositions are the result of a general-purpose learning mechanism, the evolutionary function of the learning mechanism does not deliver specific functions for the newly learnt behaviours. This chapter argues that behaviours of an individual organism can acquire functions as a result of its interaction with its environment, independently of which evolutionary functions it has. Furthermore, swampman suggests that we can representationally explain the behaviour of complex organisms, interacting with an environment, irrespective of their evolutionary history. Neither of these considerations is a decisive objection to teleosemantics' claim that the functions relevant to representation must ultimately find a basis in natural selection. However, they do motivate us to look for a way of specifying the task being performed by a system, for the purpose of varitel semantics, that does not depend on its deep evolutionary history.

My account of function will combine two elements. These broadly correspond to the two strands of Aristotelian teleology: a functional outcome is a natural occurrence that comes about always or for the most part, and that is for the sake of something (Shields 2013). The first corresponds to my robust outcomes: an organism is disposed to achieve the outcome in a range of circumstances and tends to pursue the outcome in the face of obstacles until it is achieved. The second strand is *consequence etiology*: an organism produces an outcome because of the consequences that flow from it. How can behaviour be caused because of its consequences? That in turn is naturalistically explicable if the outcome has been the target of some stabilizing process: the outcome is caused now partly because of consequences of producing the same type of outcome in the past.

Rather than choosing between these two elements, as most philosophical theories of function do, my account combines them (§3.2). Robust outcome functions (§3.3) combine with stabilized functions (§3.4) to form task functions (§3.5), which are the functions-for-representation which I argue are a good basis for fixing content. Section 3.6 explains how task functions give representational content its explanatory purchase and do so in a way that need not depend on evolutionary history. Finally, §3.7 briefly compares some rival accounts of function in the literature.

3.2 A Natural Cluster Underpins a Proprietary Explanatory Role

Humans and other animals are paradigmatic representation-using systems. Animal behaviour achieves a range of outcomes robustly. Complex internal workings (engaging representations) are involved in doing so. Those outcomes often contribute to survival and/or reproduction. And a consequence etiology applies: an animal has the disposition to produce these outcomes partly because outcomes of the same type have been produced in the past, when they contributed to survival of the individual, or were the targets of learning, or of natural selection. That is, they have been the target of stabilizing processes. The clustering of a certain kind of causation by internal workings with robustness and stabilization underpins the explanatory purchase of representational explanation.

The cluster exists for a reason. When robustness is not due to external constraints, the disposition to produce outcomes robustly is not usually possessed by accident. Often a stabilizing process has been responsible for the system acquiring robust outcome functions. An example that does not involve representations is sex determination. Since they produce an important outcome, mechanisms of sex determination have been the target of natural selection. A variety of backup mechanisms have evolved to ensure that the suite of traits needed to be a male, say, reliably come on stream together. Natural selection has made the outcome robust.

The most basic robustness tactic which evolution has hit on is survival itself. Survival of an individual organism is survival of its behavioural dispositions. Death of an organism is a form of non-robustness of all of its behavioural dispositions. It is no accident that producing outcomes robustly goes along with surviving, nor that robustly produced outcomes tend to contribute to the survival of the organism. One might object here that natural selection is really only about reproduction. Survival of the individual is at best subsidiary, and many traits are directed at reproduction in a way that compromises survival (Griffiths 2009). That is obviously correct: not all adaptations are survival-conducive. However, our project is not to define the ambit of natural selection, but to look for patterns in nature. From that perspective, it is striking that so much behaviour in the animal kingdom is conducive to survival. That is because it has contributed to reproduction by contributing to survival. Because that way of being selected is so widespread, biologists typically conceive of natural selection in terms of contribution to reproduction *and* survival. Natural selection has given us a huge array of complex systems that maintain themselves in a state that is out of equilibrium with their environment (homeostasis) and act in ways that promote their own survival.

Evolution's other great robustness trick, exemplified in animal behaviour, is learning. Learning when a particular behaviour promotes survival is a way of making survival more robust. Learning new ways to behave generates new routes by which outcomes can be produced—general outcomes like survival and reproduction, and also more specific outcomes like avoiding a predator or getting a foodstuff. Learning new circumstances in which to perform, or new routes to generating, a behavioural outcome is a way of making that outcome more robust. Learning, like evolution, is a stabilizing process by which behavioural outcomes come to be robustly produced.

These three stabilizing processes—natural selection, learning, and contribution to survival—are at work throughout the animal kingdom. Each is a way that production of an outcome in the past contributes to raising the chance that an outcome of the same type is produced again. That is, each is a form of consequence etiology. They are all processes, on different timescales, which make production of an outcome of a particular type more likely. Furthermore, both learning and evolution are ways that a particular behaviour can come to be produced more robustly: learning from feedback allows an organism to overcome obstacles or learn new routes to producing an outcome; and evolution can canalize a selected outcome so it is produced more robustly. Contributing to survival is not on its own a mechanism by which behaviours come to be produced more robustly, but for a biological organism, which is a complex out-of-equilibrium self-producing system (§3.4b below), producing outputs that contribute to its own persistence is an indispensable prerequisite for survival, which as we have seen is biology's most basic robustness trick. These are the reasons why robust outcomes tend to have been the target of one or more of these stabilizing processes. Robustness and stabilization come together in our cluster.

For example, one robust outcome function observed in the behaviour of mountain chickadees (*Poecile gambeli*) is their disposition to return to a previously cached item of food, doing so in a variety of conditions, from a variety of starting locations, and when the food is hidden in different ways (Pravosudov and Clayton 2001). Consider an individual chickadee, call her Jayla. Jayla's having retrieved cached food in the past is a cause of her survival. So, when she retrieves a food item now, obtaining cached food in the past is a contributory cause. Obtaining food is such a basic need that it is also the target of several learning mechanisms. Jayla's having this form of behavioural disposition now is partly explained by the outcomes it has produced in Jayla's past, namely obtaining food. So, obtaining cached food is, on the basis of learning, a stabilized function of Jayla's behaviour. Furthermore, learning in this way has doubtless been the result of natural selection. Natural selection explains why chickadees are disposed to return to cached food locations and can do so robustly, doubtless in part through explaining why various learning mechanisms directed at getting food have been selected. So, natural selection partly accounts for the instance of these dispositions we find in this individual, Jayla, around today. This is a paradigm case: all three stabilization processes have been at work. Each separately is a basis on which the outcome of getting food is a stabilized function of the bird's behaviour. Thus, having a stabilized function does not depend on having an evolutionary history (§3.6 below). Nor need all three stabilization processes be pulling in the same direction, as they are in this paradigm case.

In sum, there are natural reasons why, in biological organisms, robust outcome functions also tend to be stabilized functions. These come together to constitute *task functions*. It is usual to talk of entities having functions, the function to produce a certain output or cause a certain outcome. The outcomes so produced are also sometimes described as functions. It will be convenient for us to adopt that (slightly strained) terminology. So, task functions are outputs produced by a system. A type of output counts as a task function if it is robust (§3.3) and has been stabilized (§3.4). Outcomes can also be robust as a result of intentional design. That forms a further, alternative basis of task functions (§3.5).

Noting that robustness and stabilization converge still leaves open the question of how an organism manages to achieve outcomes robustly. What is the synchronic mechanism by which those outcomes are produced, and produced robustly in the face of variation in conditions encountered when they are initiated and while they are being executed? What was the synchronic mechanism that keyed those behaviours into conditions in which they were stabilized through survival, learning, and/or natural selection?

Task functions need not be generated by representations of, for example, conditions, goals, or targets. Developmental outcomes can be robust in virtue of a collection of parallel and backup mechanisms without any representations being involved. Nevertheless, in many cases there is an internal-components explanation of how the system achieves its task functions, an explanation that falls within our overall framework for representational content.[1] There are internal components which stand in exploitable relations to aspects of the environment that are relevant to achieving an outcome (a task function), where an internal process performed over vehicles with those properties constitutes an algorithm for achieving the distally characterized outcome successfully in a context-sensitive way.[2] That is to say, the third element in the natural cluster is having the kind of internal organization that is characteristic of being a representational system of the kind we have been discussing. This third element of the cluster is made more precise in other chapters—in particular, Chapters 4 and 5 specify the kinds of algorithm involved.

In short, we can observe that three features tend to cluster together: producing outcomes robustly, those outcomes having been stabilized, and their being produced by a mechanism in which internal components stand in exploitable relations to relevant features of the environment (see Figure 3.1). It is the existence of this clustering that constitutes the internal components as being representations and gives representational explanation its distinctive explanatory bite. This collection of real patterns allows us to make a rich set of inferences when we recognize a system's representational properties. When we come across instances of this cluster, a whole new explanatory scheme comes into a play, a scheme which supports a host of defeasible inferences—inferences for example about ways of acquiring and weighing sources of information, of building constancy mechanisms, and of processing information optimally, to give just three examples from the host of findings catalogued by psychology, information theory and the other cognitive sciences. On one reading of the homeostatic property cluster view of natural kinds (Boyd 1991), having representational content in accordance with this sufficient condition is a natural kind.[3] Finding a system of this special

[1] The idea that the functions or capacities of a system can be explained through causal decomposition is familiar from Cummins (1984). Unlike our task functions, which are outputs of the system of interest, Cummins functions are activities of components, each playing its role in one of these causal decompositions. Any capacity of a system is a candidate for analysis, so Cummins functions are very liberal. Without a principled way to identify privileged capacities of the system, the resulting theory of content is correspondingly liberal (Cummins 1989, 1996), contra our desideratum.

[2] Neander (2017) advances a theory of content based on contributions of components in a functional decomposition. Unlike Cummins, Neander does identify privileged capacities that call for such an explanation (e.g. the toad's prey-capture capacity). Contents are fixed directly by teleofunctions of components, e.g. a function to respond to small dark moving objects of a delimited kind in the environment, see §6.2h.

[3] With Boyd, I reject the need for an underlying essence that explains why these features go together. (The explanation is the one we have seen.) However, I don't take this core set of features to be flexible. My account requires all three features to be present. The many other properties that often go along with being a representer are however more open-ended and flexible, as with other homeostatic property cluster views of kinds. See also §8.2.

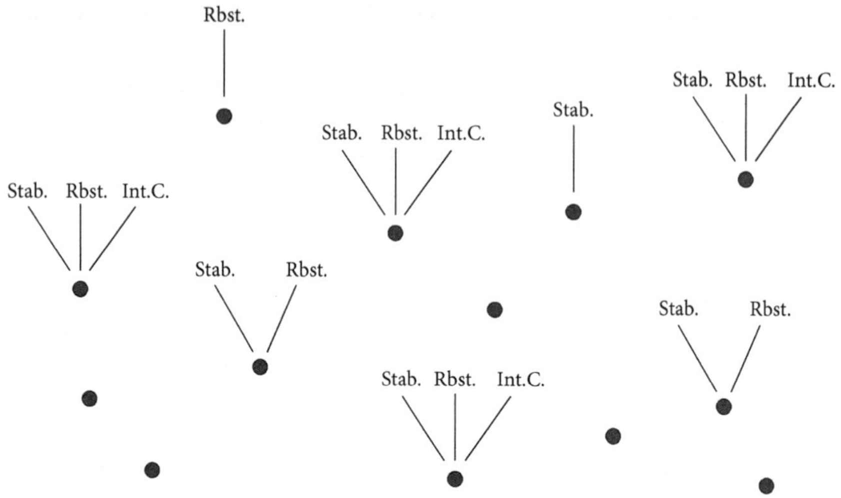

Figure 3.1 Outcomes produced by organisms/systems (black circles) can be robust (Rbst.), stabilized (Stab.) and/or produced as a result of the interaction of internal components bearing exploitable relations to the environment (Int.C.). Outcomes can have subsets of these features (labelled), or none (unlabelled black circles), however these features tend to cluster together, and do so for a natural reason (see text).

type tells us a lot about it, allowing us to predict and explain it in ways that would be unavailable or less perspicuous in non-representational terms.

The next two sections characterize the two aspects of task function more precisely, illustrated by a case study from psychology on the mechanisms of motor control. We start with robust outcome functions and then move on to precisify stabilized functions.

3.3 Robust Outcome Functions

The first requirement on task functions is that they should be robust. This section develops the relevant notion of robustness. Robust outcome functions are roughly those outcomes that result from behaviour which we humans are inclined to perceive as being goal-directed. Think of the squirrel which raids nuts from a bird feeder on a pole, crawling along a thin branch, battling with the wind, losing balance and recovering, overcoming the 'squirrel-proof' collar on the feeder, and obtaining the food. It is impossible to watch the squirrel's antics without its goal seeming obvious. The tendency to see behaviour as goal-directed can in fact be activated by appropriate movements of geometric shapes, as well as human and animal behaviour. It develops early in childhood and seems to be an important precursor to an explicit understanding of others' mental states (Abell et al. 2000, Aschersleben et al. 2008, Biro and Leslie 2007, Frith and Frith 1999, Gergely and Csibra 2003). Although we tend to see them as such, not all robustly produced behavioural outcomes depend on represented goals. For our purposes it is important to characterize robust outcome functions without presupposing that they are generated by goal-representations (or any other representations).

Motor control of reaching offers a paradigm example of robustly-produced outcomes. This is a useful case study for us because experimental work has delivered a detailed understanding of the mechanisms by which movements of our limbs are controlled subpersonally so as to reach their targets fluently. There is an online mechanism adjusting the action as it unfolds and a diachronic mechanism that tunes the online mechanism as a result of feedback. The online mechanism makes continual adjustments to the movement while the action is being executed. If the target is displaced, the trajectory of the limb is adjusted so that the finger still reaches the target (Goodale et al. 1986, Schindler et al. 2004, Milner and Goodale 2006). Those adjustments are made even when the target is shifted surreptitiously during a saccade, showing that conscious recognition that the target has been displaced need not be involved in this form of control (Fourneret and Jeannerod 1998, see Figure 3.2).

The diachronic mechanism tunes the online system so that it remains effective. Subjects fitted with prismatic goggles that shift all visual input 15 degrees to the left initially make mistakes when trying to touch a target, reaching nearly 15 degrees to the right. Over a series of trials their dispositions adjust so that they reach the target again (Redding and Wallace 1997, Clower et al. 1996). Online guidance control remains in place, with the in-flight adjustments now being appropriate to the new set-up. When the goggles are removed, an error in the opposite direction is observed and adaptation in the reverse direction begins. Similar adjustments over time occur if there is interference at the output side by having subjects make their actions in an artificial force field (Thoroughman and Shadmehr 2000). This mechanism of adaptation recalibrates our reaching dispositions as we change and grow. Patients with damage to the cerebellum exhibit online guidance control of reaching but their behaviour does not adapt to prism goggles or an artificial force field (Smith and Shadmehr 2005, Bastian 2006).

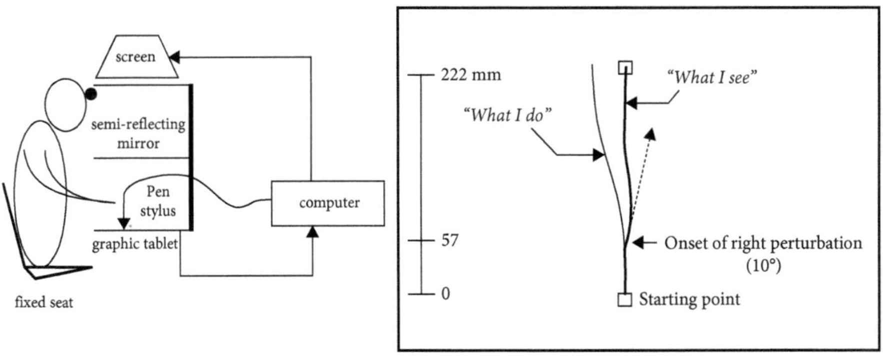

Figure 3.2 The task from Fourneret and Jeannerod (1998). Subjects adjust their reaching trajectory during action execution even when the target is moved surreptitiously during a saccade.

Motor control illustrates two key features of robust outcome functions: (i) the same distal outcome is produced in response to a variety of different inputs to the system; and (ii) the outcome is produced successfully across an array of relevant external conditions. This corresponds to the way Ernst Nagel characterized goal-directedness (his 'system property' view: Nagel 1977, pp. 271–6; crediting Sommerhoff 1950; see also Bedau 1992). Nagel separated out two kinds of variation across which the same outcome is produced or pursued: variations in initial conditions, and perturbations occurring during action execution. In many cases a perturbation can be considered as simply producing a new initial condition from which the organism may be able to reach the same goal. If our squirrel falls off the branch during its approach to the feeder, then its location on the ground is a new condition in which it will still be able to pursue the goal of getting the nuts. Other perturbations are external conditions that would prevent the system from reaching its goal, like the wind encountered when the squirrel is balancing along a fence. We can simply treat external circumstances encountered at the outset and during execution as specifying a complex condition under which the system may or may not reach a given target. Robust outcome functions are successful across a variety of such conditions, and the system is disposed to produce such outcomes in response to a variety of different inputs.

Some authors have proposed a further requirement for behaviour to count as goal directed: that the organism should bring about the outcome robustly by doing different things in different circumstances (Walsh 2012: selecting actions that are goal-conducive from an available repertoire). Should that requirement be built into our account of robustness? It is indeed a feature of motor control. Online guidance means that different sequences of motor output are deployed depending on the obstacles and disturbances encountered during execution (Schindler et al. 2004). The previous paragraph advanced a more minimal condition. Whether the organism produces the outcome must be input-sensitive, and it must do so in response to different inputs. Should we also require that it should use a range of different means?

Counting against the stronger requirement, it is common for natural selection to result in a cover-all strategy, where producing the outcome is sensitive to relevant external circumstances, but the outcome is produced by just one means. For example, one way of getting a peg into a hole is to grab it with a rubbery arm that shakes indiscriminately rather than targeting the hole.[4] For a biological example, consider a plant that pursues a cover-all strategy to get a seed into a light gap in the forest: it distributes seeds indiscriminately in all directions. Natural selection will typically make that behaviour sensitive to a variety of different cues about the season, so that the behaviour is produced at an appropriate time. But the outcome is not brought about via a variety of behavioural outputs. The argument that stabilization and robustness are linked in a

[4] Thanks to Andy Clark for the example.

natural cluster extends to such cases, so we should not require robust outcome functions to be produced by a repertoire of different means.

Notice that I did not say that the organism has to be targeting its behaviour on an object, still less that input sensitivity must be a matter of sensitivity to features of an object, like tracking its location. Cybernetic accounts of goal-directedness are modelled on control systems that achieve a goal by interacting with a target (our simple motor control case is like that). Cybernetic accounts do not extend easily to behaviour when there is no target object present, for example foraging for an absent foodstuff (Scheffler 1959). Our account of robust outcome functions has no such limitation. Nor is automatic behaviour excluded by this account. Automatic and stereotyped behaviour, like the frog's tongue protrusion in response to very specific fly-like visual stimuli, can in principle count, provided it is produced in response to different inputs and the behavioural outcome is achieved in a range of different external circumstances.

Nevertheless, not every kind of behaviour that is 'robust' in a pre-theoretic sense will qualify as a robust outcome function. A ball that simply shakes itself would reach the bottom of a rough shallow crater from many different initial positions. But it is not sensitive to inputs, either in when it produces the shaking behaviour nor in what kind of output it produces. The system is not in any way adapting its behaviour to its circumstances. Shaking indiscriminately in all circumstances is not the kind of behaviour that calls for representational explanation.

Taking stock, we have arrived at the following definition of when an output F counts as a robust outcome function produced by a system S. F is a type of output. S can be an individual system or a lineage of systems. In the second case 'S' picks out systems typed by the property in virtue of which they fall in the same lineage (e.g. being members of the same species). Recall that calling an output F a function is shorthand for S having the function to produce F (in certain circumstances).

Robust Outcome Function
An output F from a system S is a *robust outcome function* of S
iff
(i) S produces F in response to a range of different inputs; and
(ii) S produces F in a range of different relevant external conditions.[5]

'Output' is a neutral term that covers bodily movements, actions, and consequences of actions.[6] As I use the terms, bodily movements can be characterized by purely intrinsic properties of the system, for example moving the eyes 12 degrees to the right is a bodily movement. Actions can be and usually are world-involving; for example, pulling a lever or moving to a particular location. Actions have consequences in the world which

[5] 'S produces F' must be true with some nomological modal force. I remain neutral on whether this should be cashed out in terms of dispositions, capacities or in some other way.

[6] It could in principle cover any kind of effect, e.g. releasing a hormone, although movement is involved in all the cases we will consider.

may or may not be further actions. Getting a pool ball into a pocket is an action; winning £50 as a result is a consequence. All of these are 'outputs' caused by the agent and could qualify as robust outcome functions.

For condition (i), we need to look at the facts of a particular case to assess what counts as a different input. They have to be differences that the system is sensitive to in some way (e.g. an undetectable difference cannot count). A generalization that follows from how a mechanism deals with one type of input is not sufficient either. For example, a mechanism that triggers an internal state R when a temperature of 20°C is detected might, without further elaboration, do the same at 19.5°C and 20.5°C. Evolutionary pressures could select for this kind of stimulus generalization. Nevertheless, these values would not count as different inputs. They specify the range of values that count as an input of the same type for this kind of mechanism. If, on the other hand, R is triggered by a temperature of 20°C and also an increase in light levels, then those do count as different inputs.

The idea of different relevant external conditions in (ii) also needs careful handling. A different alignment of the planets is a different external condition but is not (usually) relevant to whether an outcome can be successfully produced. Relevant conditions are those that would impact on the system's ability to achieve an outcome or would affect whether the outcome is likely to be successful. In the seed-scattering example, the difference in where the nearest light gap is found, and hence where the seedling germinates, is a difference in a relevant external condition.

3.4 Stabilized Functions: Three Types

(a) Consequence etiology in general, and natural selection

The second element in our cluster is the category of stabilized functions. They correspond broadly to the second aspect of Aristotelian teleology: the idea that teleological outputs are produced because they lead to good consequences. In discussing the cluster (§3.2) I argued that robust outcomes tend to go along with being the target of natural selection and/or of learning, and/or with contributing to the persistence of individual organisms. This section spells out those conditions more precisely. Taken in the alternative, they define our category of stabilized functions.

How can an output be generated because of the consequences it will produce? Of course, an agent can do that. But agency presupposes intentionality. For a long time it was not clear how to account for teleological causation without presupposing intentionality. Darwin famously showed that there is no mystery. An output is generated because of the good effect it will produce when, in an organism's evolutionary history, outputs of that type have contributed to survival or reproduction. In that case the organism is producing this output now in part because of effects that the same type of output has had in the (evolutionary) past. Larry Wright generalized that idea: F is a function of S just in case (a) F is a consequence of S's being there; and (b) S is there because it does F (Wright 1973). Wright's definition covers processes

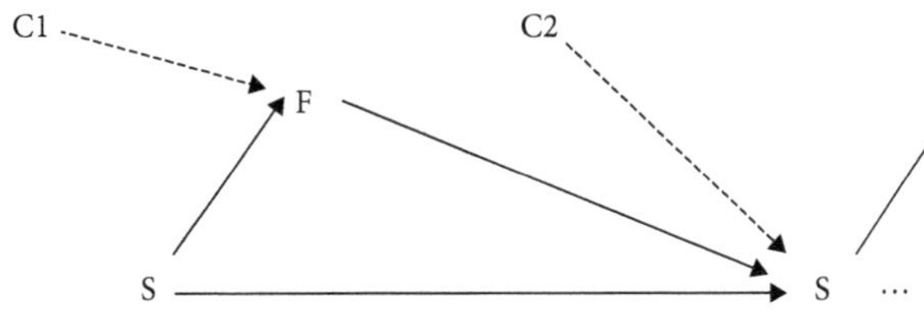

Figure 3.3 The most general treatment of consequence etiology. An output F occurs because a system S exists (right-hand side); and S exists because it, or its ancestors, have produced F in the past (left-hand side). Both causal processes may depend on certain environmental conditions obtaining (C1, C2).

like feedback-based learning operating within the life of an individual organism as well as processes like evolution by natural selection operating over lineages of organisms.[7] It applies to any process where outputs in the past have had consequences that explain the current existence of a system disposed to produce outputs of the same type. I use 'consequence etiology' to cover any output which satisfies Wright's definition (see Figure 3.3).

Wright's definition has long faced the objection that it draws the category of function too broadly (Boorse 1976). For our purposes that is problematic because the definition is much broader than the kinds of stabilizing processes found in our natural cluster. It applies to a small rock that keeps its place on the river bed by holding a larger rock in place above it in a stream; also to a leaky gas hose that keeps on emitting gas because the gas poisons every person that comes near enough to fix it. Contribution to survival of an organism is perhaps the most widely applicable kind of stabilization in our natural cluster, but even that is a special case of Wright's formula. It calls for an organism, one that acts to maintain or promote its survival in the face of changes to internal and external conditions.

Our task then is to delineate the class of stabilized functions in a narrower way than Wright, so as to coincide with the natural cluster that underpins representational explanation. Since any single cover-all condition like Wright's is liable to over-generate, I will adopt a disjunctive definition of stabilized functions. Evolution by natural selection is the first case. It is a well-understood basis for stabilized functions. I intend it to extend to cases where selection stabilizes the presence of a trait in a population but has not gone to fixation; also to selection on culturally transmitted traits. The next two subsections focus in turn on the other two kinds of consequence etiology that show up in our natural cluster: contribution to persistence of an organism; and learning with feedback.

(b) *Persistence of organisms*

The most ubiquitous way that natural selection has made outcomes robust is by inventing the organism: a complex system that is separated from and markedly out of

[7] E.g. it covers all the various kinds of dynamics studied in Skyrms-Lewis signalling games: replicator dynamics (with and without mutation), simple reinforcement learning, Roth-Erev reinforcement, Bush-Mosteller reinforcement, etc. (Skyrms 2010).

equilibrium with its surrounding environment, and which continually creates the conditions for its own persistence in that state.[8] By continuing to exist, organisms are able to continue to produce the types of outputs they have produced in the past, enabling robustness.

Philosophers have put forward several accounts of biological function in terms of contribution to persistence: staying alive (Wouters 1995, 2007), self-reproduction (Schlosser 1998), active self-maintenance (Edin 2008), or maintaining a differentiated organized system (Mossio et al. 2009). Christensen and Bickhard's account of functions is in this spirit (Christensen and Bickhard 2002). According to them the task towards which functions are directed is a system's capacity to generate the conditions for its own persistence when it is out of equilibrium with its surrounding environment. Functions of components of a system are Cummins-style contributions to this overall capacity.

Our stabilized functions are outputs of a whole system, rather than of components,[9] but they contribute to persistence in something like this fashion. Rather than starting with difficult concepts like self-maintenance and being out-of-equilibrium in the appropriate way, we can focus on the kind of persistence that figures in our cluster—that is, the persistence of organisms. Organisms are a special kind of self-maintaining system. They resist the tendency to disorder by maintaining a boundary, moving energy across it, and continually rebuilding themselves to keep themselves in an improbable state of differentiated organization. Godfrey-Smith uses the term 'self-production' to distinguish organisms from other self-maintaining systems like a car that monitors its states and fixes some problems (Godfrey-Smith 2016, following 'autopoiesis': Maturana and Varela 1980). Organisms are also self-producing in a stronger sense than we find in cases like the leaky gas hose and the rock on the river bed. An account of what it takes to be an organism would open up debates about equilibrium, self-maintenance, and self-production which would distract us from our enquiry, so my definition will help itself to *organism* as a biological category. It is contribution to the persistence of an organism that should count as a stabilized function for our purposes.[10]

Chemotaxis in *E. coli* bacteria is a good illustration of the way behaviour can contribute to the persistence of an individual organism. An individual swims in a straight line, but when it detects that the concentration of one of a number of harmful chemicals is increasing, it performs a random 'tumble' to a new direction of travel (Berg and Brown 1972). The effect of this behaviour is to take the bacterium away from dangerous chemicals (often enough), thereby contributing to its persistence. Moving away from harmful chemicals is a distal outcome of the bacterium's behaviour, an outcome that contributes to its persistence. This is a typical case where robustness of the outcome (safe location) in the face of variation in external and internal biochemical

[8] Reproduction of entities that don't count as organisms/autopoetic systems is in principle possible, although there is debate about whether there was actually such a stage in the origin of life (Martin 2005).

[9] See §3.3 (unless those components count as systems in their own right).

[10] In what follows 'persistence' is always persistence *of an organism*, even when I omit the qualification for the sake of brevity.

parameters (Alon et al. 1999) goes together with those outcomes contributing to the persistence of the individual organism.

Where an output has contributed to the persistence of an individual organism we can give a consequence etiology explanation of its current behaviour. It behaves a certain way now partly because it behaved in the same way in the past, which had consequences that kept it alive, raising the probability that there is something around now that will produce an output of the same type. We come across an instance of bacterial tumbling behaviour partly because that kind of behaviour has kept the individual bacterium alive, together with its disposition to tumbling behaviour. That is a historical rather than a forward-looking or counterfactual way of explaining behavioural outputs. Indeed, without the historical angle we would be back to the mystery of teleological causation, the mystery of how it is possible to explain a cause in terms of the type of effect it is likely to produce (without appealing to intentionality in the causal agent).

When an occurrence of F contributes to the persistence of an organism S, its effect is not specific to F. It raises the probability that any of S's outputs will be produced (since S is still around to produce them all). That is unlike natural selection, which increases the probability of organisms producing F rather than alternatives; or feedback-based learning, which specifically raises the probability of S producing F in appropriate circumstances. Furthermore, in an organism capable of feedback-based learning, contribution to persistence has the effect of keeping an organism with the disposition to F around for long enough for learning to make F-production more robust, improving its discrimination about when to produce F or acquiring new means for bringing it about. Persistence is then an indirect route to making an outcome F robust.

(c) Learning with feedback

Returning to our motor control example, the way reaching adapts when subjects put on prism goggles illustrates the importance of learning in producing robustness. Reaching behaviour studied in non-human animals often leads to outcomes that contribute directly to the persistence of the animal performing the experiment. A macaque receives food or juice as a result of where it reaches or moves its eyes, which contributes directly to the persistence of that individual organism (Kiani and Shadlen 2009, Chestek et al. 2007). Human subjects are generally rewarded with money or course credit rather than food. In that case the outcomes produced don't directly explain the persistence of the individual organism.

Outcomes do, however, explain why a given behavioural tendency arises or persists in an individual. For example, a person in a reinforcement learning experiment might learn to press the 'F' key on a keyboard in response to some arbitrary image A and the 'J' key in response to another image B. Those behaviours are reinforced because the subject is given points that will turn into a monetary reward at the end of the experiment. If we focus on the disposition to press the 'F' key in response to image A, then an account of why that behavioural disposition exists in the individual mentions the outcomes that have been elicited by pressing the 'F' key in the recent past. Learning can

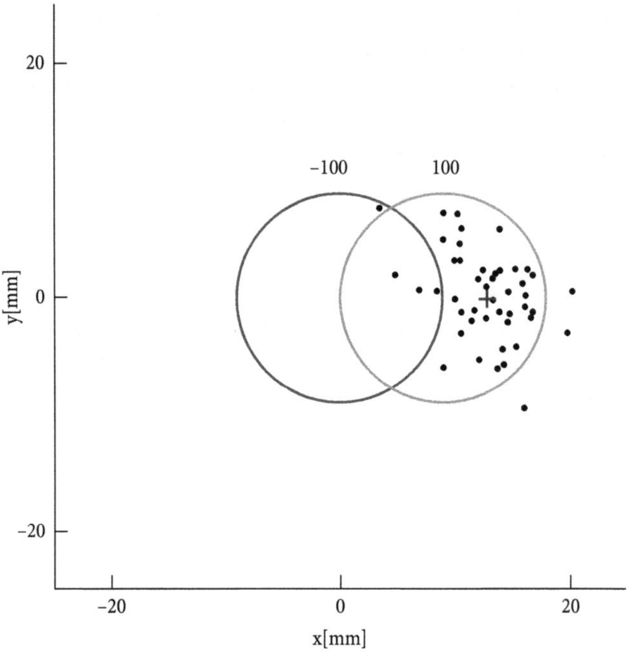

Figure 3.4 The rapid reaching task from Wolpert and Landy (2012). Subjects gained 100 points for touching the screen within the right-hand circle (shown to subjects in green) and lost 100 points for touching in the left-hand circle (shown to subjects in red). Touching in the overlap therefore produced zero points. People mainly touch within the most rewarding area because they learn from feedback how to reach (observing previous outcomes).

also explain the robustness of a behavioural disposition; for example, the ability repeatedly to touch a screen within a small target area, across small variations in the initial conditions, and in the face of noise in the perceptual and motor systems (Wolpert and Landy 2012; see Figure 3.4). There is of course also a learning explanation of the macaque's reaching behaviour. It is stabilized both by learning and by contribution to persistence.

Isn't flexibility the converse of robustness? Learning is an interesting case because it shows how flexibility is important for robustness. We often find in biology that keeping some properties constant calls for sensitive flexibility in others. We see this in the way that motor control is constantly being retuned as the system's input and output properties change (optical properties, weight of the limbs). Learning allows for plasticitiy in the circumstances in which, and means by which, an outcome in produced—leading stabilized outcomes to be produced with greater robustness.

Learnt behaviours have evolutionary functions that derive from the function of the learning mechanism (Millikan 1984). Humans readily learn to recognize conspecifics by their faces. Human infants look preferentially at faces, which allows them to learn the statistical patterns that are indicative of face identity (Johnson et al. 1991). If we suppose for a moment that no social feedback is involved, then the reason the infant acquires a new behavioural disposition—for example, to track a new person A as they come and go—does not depend on any feedback the individual infant has received. The function of the mechanism is indeed to track person A, but that is an evolutionary function, deriving from the (plausible) evolutionary function of the learning mechanism, namely to track conspecifics by their faces. This is a case where evolutionary functions do deliver quite specific stabilized functions for the products of learning.

In other cases derived functions are much less specific. Classical conditioning is a very general learning principle. It enables organisms to reidentify statistical patterns in the inputs they receive. When an association has been learned, what is it supposed to track? The evolutionary function of the learning mechanism only tells us something very general. Its function is to track something useful that correlates with patterns in the input. Once a new association is put to use to condition behaviour, if that behaviour is stabilized then feedback-based learning may underpin a much more specific function, as we will see in a moment. But before being connected up to behaviour, the functions of a new association derive only from the evolutionary function of classical conditioning and are highly indeterminate. Basic sensitization, where a response is attenuated when a stimulus is repeated, is another case where the mechanism of behavioural plasticity has only a very general-purpose evolutionary function.

When an organism's behavioural dispositions are modulated by feedback, learning underpins stabilization directly, irrespective of any evolutionary function. Feedback need not be in the form of a commodity that has an evolutionary function (a primary reinforcer). People will shape their behaviour to feedback in the form of money, or the promise of money, or tokens that will be exchanged for money; also for positive social feedback; and so on. A stabilization-based explanation need not descend to an explanation of why monetary feedback stabilizes behavioural dispositions. If an agent's behavioural dispositions are in fact stabilized by a variety of outcomes O_i, then we can explain a current behavioural disposition (e.g. to touch the region inside the green circle on the computer screen) by the fact that outputs of this sort in the recent past produced one such outcome, O_1 say. It is then a further question as to why O_1 reinforces behavioural dispositions in that agent.[11] An answer to that question need not form part of a stabilization-based explanation of why the agent has a given behavioural disposition now (e.g. to touch inside the green circle).

Like natural selection, reinforcement can lead an organism to produce O more robustly by better detecting the circumstances in which its behaviour is likely to produce O; by adopting new means for producing O in new circumstances; or by increasing the robustness with which it can produce a particular means to producing O. Learning is more sophisticated than natural selection in some respects. One-shot learning is

[11] That need not be because the organism can represent the reinforcer. Nor does the learning-based explanation of an organism's behavioural dispositions presuppose that learning depended on representations (of reinforcers or outcomes).

possible in some cases. Then a single episode in the past explains why an individual has a behavioural disposition now. Nearby outcomes can be reinforcing. Where O is the target of learning, achieving an outcome that is close to O (along some relevant dimension) can make it more likely that the organism will achieve O on the next occasion. That is, outcomes that are closely related to O can contribute to likelihood that O will occur in the future. Where an outcome comes in degrees, like the quantity of juice received, the organism may shape its behaviour so as to increase the quantity delivered. Negative reinforcement is also common, for example a rat forced to swim in a Morris water maze will learn how to behave so that it has to swim for less time in the future. So, the stabilized function is getting to a submerged platform (O_1) and the feedback which explains its stabilization is the unpleasant effect of not reaching the platform ($\neg O_1$). In these two cases, it is not producing O itself but producing outcomes closely related to O that has contributed systematically to the organism's disposition to achieve O.

Learning by imitation is an interesting case. It takes several forms. Sometimes it is driven by social feedback; for instance, that people smile or give other signs of approval. That is a case of reinforcement and fits into the characterization we have just given. Or the learning may occur because the individual performs the behaviour and receives some other kind of reinforcing feedback, like food or warmth. In other cases people may acquire a behavioural disposition without feedback, just because they see others perform it. That disposition will not then have been stabilized through feedback-based learning, but there may well be another stabilization-based explanation; for example, the behaviour which is transmitted may have been stabilized in the person's lineage or social group through cultural evolution.

It would take us too far astray to catalogue all the types of learning and to give an account of what characterizes the different kinds. For our purposes it is enough to point to the category of feedback-based learning, as used in the behavioural sciences, and to note that it is a strong form of stabilization that tends to go with robustness in our natural cluster.

(d) A 'very modern history' theory of functions

This section constructs my notion of stabilized function out of natural selection, learning, and contribution to persistence, and defends its historical character.

It would be handy if we could treat stabilization synchronically, on the model of forces that are holding a system in place. But dispositions an organism could exercise are not like forces or other outputs that are operating continuously. Our stabilized functions are not like the kinematic equilibria studied in physics. That makes it tempting to adopt a counterfactual or forward-looking approach. Stabilized functions would then be outcomes that would be stabilized were they to be produced, or that are likely to be stabilized in the future.

The difficulty is that it is a very open-ended matter whether an output would contribute to the persistence of an organism, or would be stabilized by feedback-based learning, or would promote reproductive fitness. All outcomes that would contribute to the persistence of an individual would count as being amongst its stabilized functions. However, whether an outcome will contribute to persistence is a notoriously open-ended matter. It depends heavily on the context. And within a context, whether a behaviour will in fact be stabilized will depend upon accidental features of the process that ensues. Outputs that seem very unlikely to contribute to persistence might end up doing so through a series of compensatory accidents (as happens to the cartoon character Mr Magoo). Without some other constraints, there are just too many effects that would be stabilized in some circumstance or other, hence too many functions. Facts about what could contribute to persistence are much more open-ended than historical facts about what has actually contributed to the persistence of an individual organism. The same is true for natural selection and for learning.

A second strong reason not to rely on a forward-looking form of stabilization is that such functions are of the wrong kind to figure in causal explanations. Recall the mystery of teleological causation, namely of understanding how a good effect could 'draw out' a cause suited to producing it. Wright's way of making that un-mysterious, and Darwin's, is to point to consequence etiology, in which functions are a matter of the effects that such outcomes have produced in the past. If we seek to explain why a system produced an outcome O, it is unilluminating to cite the fact that O is likely to be stabilized in the future (i.e. to cite a future-directed function). Functions based on stabilization history can figure in an explanation of why outcomes are produced; forward-looking functions cannot (not straightforwardly). A historically based approach to function thus has better credentials to figure in casual explanations than a forward-looking approach to function does. Any explanatory purchase of forward-looking functions would proceed through a historical generalization—that they tend to have been the result of some stabilization process. Furthermore, our functions need to connect with the natural cluster that underpins representational explanation. It is actual historical processes of stabilization that appear in that cluster.

What is relevant for our purposes, then, is contribution to stabilization through: natural selection over a lineage of systems, learning within an individual system, or persistence of an individual organism. Teleosemantics standardly appeals to the first two (although with some problems with the way learning is incorporated). I expand the category to accept the widespread suggestion (e.g. Christensen and Bickhard 2002) that functions can be a matter of contribution to the persistence of self-producing systems (for our purposes, organisms). I follow Godfrey-Smith's insight that appealing to actual causal history is the right way to cut down on the problematic liberality of forward-looking accounts of function (Godfrey-Smith 1994b). Godfrey-Smith calls his appeal to the most recent evolutionary function of a trait a 'modern history' theory of function. We could then call ours, which includes the recent learning and persistence history of an individual organism, a 'very modern history' theory of function. These functions can arise just from the history of an individual organism, including very recent learning and contributions to its persistence, irrespective of any history of selection.

Stabilized Function
An output F from a system S is a *stabilized function* of S
iff
producing F has been systematically stabilized:
(i) by contributing directly to the evolutionary success of systems S producing F; or
(ii) by contributing through learning[12] to S's diposition to produce F; or
(iii) where S is an organism, by contributing directly to the persistence of S.

The evolutionary condition is deliberately drawn so as to cover cases of cultural transmission, which may have been important in human cognitive evolution and thus in generating representational content in many aspects of human psychological systems (Sterelny 2015). It also covers cases where selection has been operative but has not gone to fixation.

A system's behaviour will generally result in a causal chain of outputs, which can vary in robustness along the chain. Stabilization homes in on only one or a small number of steps in this chain. When a macaque moves its arm to pick up a grape, getting the grape and moving the arm both make a causal contribution to that form of behaviour being stabilized, and to the individual persisting, but only getting the grape does so directly. On the other hand, idiosyncratic things may have happened in the individual's history. Where a behavioural episode accidentally happens to produce some beneficial or reinforcing effect, and there is no systematic explanation of why that is so, then it does not even start to generate stabilized functions, even if the episode made some contribution to the persistence of the individual or the chance that it would produce a particular kind of behaviour in the future.

3.5 Task Functions

This section puts the pieces together and defines task function, which I argue is the right account of function to figure in accounts of content in our case studies. Task functions combine stabilization with robustness. There is a source of robustness which we have not yet considered, namely deliberate design. A human can design a system to perform a task; that is, to produce certain outcomes robustly in certain circumstances. Design need not involve any history of stabilization. Indeed, artefacts can be designed to produce an outcome robustly which would not be stabilized by feedback. For example, we could design a robot that would navigate to a power source and use the energy to blow itself up, with the ability to do so robustly from a variety of starting points via a variety of routes. So, we need to include design functions as an alternative to stabilized functions.

[12] As discussed in §3.4c, this is intended also to cover nearby reinforcement, where producing an outcome close to F (along a relevant dimension) accounts for S's disposition to produce F; also negative reinforcement, where the disposition to do F has been stabilized by the negative consequences that have flowed from doing things contrary to F.

Task functions based on design do not meet our criteria for naturalism. What a system has been intentionally designed to do depends on the mental states of the designer, so this is not a non-semantic, non-mental source of functions. It does not form part of our account of where underived content comes from. Nevertheless, it is worth recognizing the products of design as having task functions since they pattern with the other cases, before setting aside design in order to focus on the underived cases.

The kind of design we include is where a person has designed a system to produce certain behavioural outputs. Another way content can arise derivatively is more direct: a person can intend a representation to have a certain content. A sentence may mean what its writer intends it to mean. A computer database may represent what its programmer intends it to represent. Directly derivative content does not depend on task functions at all. It comes directly from the user's intentions or beliefs about what a vehicle represents, so our definition of task functions does not cover these cases.

So, an output is a task function if it is robustly produced and is the result of one of the three stabilizing processes discussed above or intentional design:[13]

Task Function
An output F from a system S is a *task function* of S
iff
(a) F is a robust outcome function of S;
and
(b) (i) F is a stabilized function of S; or
(ii) S has been intentionally designed to produce F.

I am not suggesting this as an analysis of biological function. Some have argued that representing correctly is genuinely normative, and that biological function is genuinely normative, and that the otherwise puzzling normativity of content can be dissolved by showing that it reduces to the normativity of biological function. It will be apparent that I am not engaged in that project. Both biological function and (subpersonal) representational content are descriptive categories (see §6.5). My definition of task function does however have several features that are familiar from biological functions. A system can have task functions that it is no longer able to perform. It can malfunction, which is different from not having a function at all. And it can produce outputs which are side effects, which regularly accompany task functions but have not been the target of stabilization or are not robustly produced.

Nor do I claim that task function is the only notion of function to which representational content can be connected (recall the pluralism). My claim is that task functions are suited to giving an account of representational content in many kinds of subpersonal psychological system. Task function is a necessary part of some sufficient conditions for content. (Since the definition of task function is disjunctive, and the

[13] Usually the contexts in which the output is produced robustly will largely coincide with the contexts in which it was stabilized.

definition of stabilized function is disjunctive, it really generates several different sufficient conditions.)

Task functions are part of a natural cluster, a real pattern in nature that I will argue gives representational content better ways of explaining behaviour than would be available otherwise (§3.6, §8.2). However, task functions can vary in ways that affect the explanatory bite of the contents they generate. This is what we should expect in biology. Robustness comes in degrees. The more robust, the more explanatory value a task function is likely to have. Similarly, stabilization comes in degrees, from powerful or long-standing stabilized functions to marginal cases; for example, evolution where there has been some selection, but no fixation or other endpoint has been reached.

Another dimension of variation lies in the various bases of stabilized functions (clauses (i)–(iii) of the definition). In a paradigmatic case an outcome like obtaining food has been the target of natural selection, and learning from feedback, and has contributed to persistence of the individual organism. But these may not line up. A rat that learns how to press a lever to deliver electrical stimulation directly to reward centres in the brain acquires new task functions (by (ii)), but ones that are evolutionarily disadvantageous (clause (i)) and do not contribute to the animal's persistence (clause (iii)). In natural cases there will normally be connections between the three stories; for example, money becomes a positive reinforcer partly because of its connection with reinforcers like social feedback for which we can give a more direct evolutionary explanation. When they dissociate, there will still be task functions, but there may be different task functions underpinned by the different clauses, and they may pull in divergent directions (see §3.7). Representational explanation will have greatest merit in the paradigmatic cases and less merit in these more marginal cases. Less paradigmatic task functions can underpin genuine representational content—they are not merely cases of 'as if' content—but if the penumbral cases were the only ones that existed in nature it would be unlikely that representational content, of the kind we have identified here, would be an important explanatory category. The marginal cases are not what makes our natural cluster explanatorily powerful, but they do get carried along for the ride.

When a property applies more widely, that is generally of explanatory benefit, but it trades off against the fact that more generally applicable properties tend to support fewer inductions. Although we can classify very many entities as *physical object under 10 kg*, falling into the category tells us little about other properties an object is likely to have—it supports few inductions. The merit of our cluster is that, as well as being found widely in nature, it supports a rich set of inductions. Robustness also gives us generality by grouping a range of different local properties together (§3.6, §8.2). A system's reaction to light and sound might be lumped together because they are both means for tracking a distal property like distance. What can look like a variety of different processes, if we considered only the local operation of the system, exhibit commonalities when treated in terms of task functions. This generality is not achieved at the cost of reduced inductive potential (as with *object under 10 kg*), since task functions key into our cluster and a rich set of world-involving inductions about the system's interaction with distal features of its environment.

3.6 How Task Functions Get Explanatory Purchase

(a) Illustrated with a toy system

In this section we look at a stylized toy system that captures some essential features of the mechanisms of motor control. It will illustrate why task functions underpin a proprietary explanatory role for content. Well-confirmed accounts of motor control appeal to a variety of interacting internal components including forward models, inverse models, and comparator circuits (Desmurget and Grafton 2000, Battaglia-Mayer et al. 2014). The most basic kind of comparator circuit compares visual or proprioceptive feedback about the location of a limb with a specification of a target location in the same code, using the discrepancy between the two to update the motor program driving the limb (Wolpert and Ghahramani 2000). In this way the limb's position is adjusted until the difference between its location and a target location is reduced to zero.

Smooth control of action also depends on making internal predictions of the likely effects of executing a motor command, and adjusting the motor command in response to discrepancies between the prediction and the target state, even before feedback from the world has been received (Wolpert et al. 2011, Bastian 2006). Since I can illustrate how representational content gets explanatory purchase even without these additional internal components, I will work with a simple toy model that only contains the first comparator circuit, the one based on external feedback. Figure 3.5 illustrates this toy system S. It moves in just one dimension, along a line. From a range of initial conditions, it will move along the line until it reaches location *T*, where it

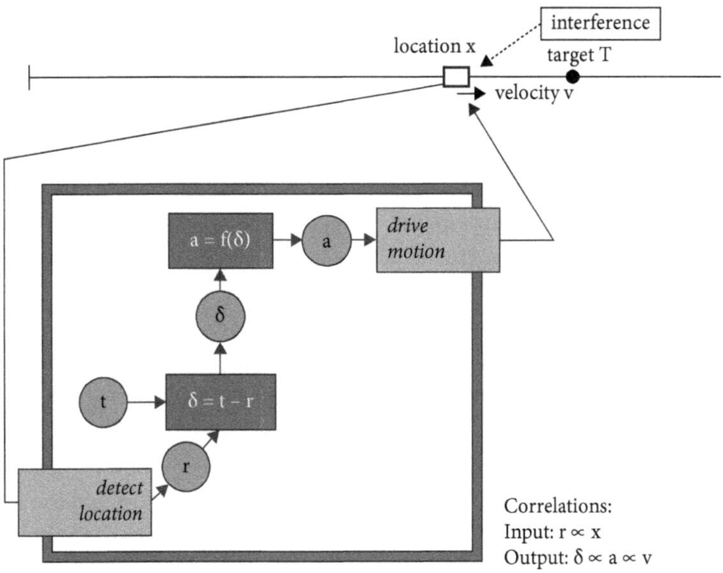

Figure 3.5 A system deploying a simple comparator mechanism.

stops. If it is blocked or displaced along the line it will continue to move towards *T* when released. Reaching *T* is a robust outcome function of the system.

We can explain how the system achieves the outcome of reaching *T* by appeal to its internal organization and the relations that those internal components bear to features of the environment. S has an internal register **r** that correlates with its distance from the origin, and another internal register **a** that correlates with the velocity of its wheels. A third internal state **δ** correlates with the distance of the system from *T*. That correlation is achieved by subtracting the activity of **r** from another fixed level of activity **t**. A monotonic transformation from this difference signal **δ** to **a** is such that the motion produced in the wheels drives S from any starting position to *T*, where it stops.

Reaching *T* is a distal outcome produced by S robustly using a variety of motor outputs—ways of changing the velocity of the wheels over time. What these different patterns of motor output share is that they all achieve the distal outcome of reaching *T*. Similarly, at input, S will reach *T* from a variety of different starting positions, and in the face of a number of ways of displacing S while it is executing its action sequence. So, reaching *T* satisfies the definition of being a robust outcome function of S. (The robustness is not very great, and so representational explanation will not deliver very much additional explanatory purchase, but the case is sufficient to illustrate the point.)

To get stabilized functions into the picture, we have to supplement the case. Suppose the system needs to recharge its batteries periodically if it is not going to stop moving altogether. There is a power source at *T*. Now, if we encounter S, moving around, with its disposition robustly to reach location *T*, there is a consequence etiology explanation of why. Reaching *T* in the recent historical past has contributed to the persistence of the system, with its disposition to reach that very location. We could also add a learning-based stabilized function. Suppose the internal state **t** is reset periodically at random, leading the system robustly to move to a new location; also suppose that when the system manages to recharge (by chance at this stage), that fixes the state of **t**. In future it will then robustly move to the recharge location *T*. Getting to *T* has then become a learning-based stabilized function. (We could add in the further capacity to adjust its dispositions over time in response to perturbations in the input and output mechanisms, as in the case of motor control—prism goggles and artificial force fields—a stronger form of learning which would produce greater behavioural robustness.) Either way, reaching *T* becomes a task function.

We now have all the elements in place to explain the system's behaviour using the standard explanatory grammar of representational explanation. S has various internal components that correlate with distal features of the environment (correlation being the relevant exploitable relation in this case). **r** correlates with S's distance from the origin and **δ** correlates with S's distance from *T*; **t** with the location of a power source. There are internal processes that transform **r** into **δ**, and **δ** into **a** and the velocity of its wheels. Given the way **r** and **δ** correlate with external features of the environment

(S's distance from the origin and from *T*, respectively), these internal transformations constitute an algorithm for performing the distally characterized task of reaching *T*.

Now consider a particular episode of behaviour: the system is displaced and moves back to *T* where it recharges. How so? Because **r** and **δ** correlated on that occasion with S's distance from the origin and from *T*, respectively. The next chapter shows how correlations are content-constituting when they explain task functions in the right way. So, the story I have just told effectively shows how the behaviour of successfully reaching *T*, the location of a power source, is explained by **r** and **δ** representing correctly. Conversely, suppose noise affects the input system and S stops at some other location *T'*. Then we can explain failure of the system to reach *T* in terms of misrepresentation by **r**. Similarly, misrepresentation by **δ** or **a** would explain unsuccessful behaviour. This pattern exemplifies the characteristic explanatory grammar of representational explanation: correct representation explains success and misrepresentation explains failure.

(b) Swamp systems

To see why this really is a case of success and failure, consider a 'swamp' system, like S, but one that has assembled itself by chance when an earthquake struck an engineer's workshop. This swamp system would be disposed to move up and down a line along the work bench, stopping when it reached some location *T*. Reaching *T* would be a robust outcome function of the system. But now consider what would happen if a random event affected internal register **t** so that S became disposed to reach a different location *T'* (and to do so robustly). Would that count as a failure, to be explained by misrepresentation? Or would it count as a success—success in achieving the system's new function of reaching *T'*, to be explained by correct representations (with different contents)? There is nothing yet in the picture that allows us to answer that question one way or another.

If we now add that there is a power source at *T* and we observe the swamp system a short time after the earthquake, when it has had a chance to move around and recharge, then we do have something in the story which underpins a notion of success and failure for that system. It is part of an explanation as to why that particular swamp system is around, with its disposition to reach *T* robustly, that it has reached *T* in the recent past, which has contributed to the persistence of S and its behavioural capacities. If noise now messes things up so that **δ** is calculated differently and it no longer reaches *T*, but robustly reaches *T'*, that is a failure, to be accounted for by misrepresentation at **δ**.

The foregoing is effectively an argument from intuition. It trades on the intuition that there is no substantial distinction to be made between success and failure in a system that has robust outcomes but no history (hence no stabilized functions from persistence, learning, or evolution). That won't do for our purposes. We could simply define robustly produced outcomes as successes and other outcomes as failures.

However, the cluster we have identified means that there is a deeper logic behind these intuitions. Robust outcomes, when not accidental or due to external constraints, are often explicable in two ways at once: both why and how they are produced. *Why* is explained by history, involving a consequence etiology, here contribution to stabilization. *How* is explained synchronically, by internal components and exploitable relations. (This is fleshed out properly in Chapters 4 and 5.) Past episodes of stabilization can explain both how robust outcomes are produced now, and why the system has a mechanism suited to producing those outcomes robustly. It is the combination of these elements which makes it the case that certain outcomes are successes and others are failures. Our intuition about the absence of success and failure in the swamp system before it has interacted with the world reflects the fact that the cluster of elements that give representational content its explanatory purchase is absent in that case.

How does this argument transfer from our toy system to organisms? Consider the motor system of a swamp macaque, produced at random by lightning striking a swamp. From the moment of creation it would have the same robust outcome functions as a regular monkey. So, if it sees a grape, the swamp monkey will grasp it and eat it. Consider also a second swamp macaque who just happens to have a robust disposition to grasp 15 degrees to the right of any grape it sees. At the moment of creation there is nothing present that underpins a substantial sense in which one swamp monkey is getting it right and the other getting it wrong. However, as soon as they have had time to interact with the world for a while, there is an important difference. In one the disposition to reach and grasp grapes has contributed causally to its disposition to behave in that way—it has been the target of learning and has contributed to persistence—in the other the disposition to reach 15 degrees to the right of grapes has not. At the moment of creation neither monkey exemplifies the cluster of properties that underpins the explanatory purchase of representational content. Correspondingly, there is no substantial sense in which either one of them is getting it right or getting it wrong. Once they interact with the world, one monkey starts to exemplify the cluster, with which comes a substantial distinction between correct and incorrect behaviour; the other does not.

These thought experiments with swamp systems are not offered as intuitive evidence in favour of my account of task functions, but rather to illustrate the consequences of the theory. Systems which are not the result of deliberate design and have no evolutionary history, but do learn systematically from feedback, would begin to acquire task functions after a short period of interaction with their environment. The same goes for organisms whose actions contribute to their own survival. This illustrates the fact that functions based on current properties of the system (robust outcome functions) and recent causal contributions to learning or persistence (stabilized functions) can underpin representational content in a way that is independent of facts about why the system was designed or its deep evolutionary history. That would be so even for a system that does in fact have an evolutionary history—the swamp system thought experiment just serves to dramatize the fact that even in natural cases, stabilized functions can arise in a way that does not depend on evolutionary history.

We can see that in the case of learning. Think of a child that learns to clap, based on social feedback from a parent. It produces outputs (ways of clapping) that make the parent smile and learns to perform the behaviour in appropriate circumstances (e.g. not at dinnertime). These outputs now have a stabilized function F: to make the parent smile. The child's behaviour has that function irrespective of any facts about evolutionary history. Non-evolution-based stabilized functions are acquired gradually as an organism interacts with its environment and receives feedback that reinforces behaviour or contributes to survival. A swamp system would have no task functions at the moment of creation, but would acquire them piecemeal, and would soon have task functions, functions keying into aspects of the environment it has interacted with. As soon as a swamp system has some interactions with the environment, then there will be an explanandum at which content-explanation can be addressed (success and failure), and the system will begin to have contentful states.

Task functions are still partly historical, so I have to bite the bullet and accept that a swamp system has no contents at the moment of creation. As I have argued, however, that is the right result. In these subpersonal systems, at the moment a swamp system is created there would be no explanandum for content-based explanation to address. However, the bullet is a lot more digestible than that confronted by standard teleosemantics, which accepts that a system with no evolutionary history would have no contents even after a long life of interacting with and learning about its environment.

We can also see how the robust outcome aspect of task functions contributes to the proprietary explanatory purchase of representational content (see also §8.2b). Because reaching *T* meets the conditions on being a robust outcome function of S, there are world-involving distal patterns involving S that are less perspicuous when we consider S's behaviour only in terms of proximal sensory stimulation and proximal motor output. The same location is reached across a variety of patterns of perceptual input. Despite the simplicity of this toy system, there are real patterns in the way S interacts with distal features of its environment that generalize across proximal inputs. (In paradigm cases there will also be generalizations across proximal outputs, with multiple motor outputs eventuating in a common distal outcome, as discussed in §3.3 above.) Explanations of S's behaviour would look more complex and disjunctive if we did not recognize those patterns.

Contrast the case of the rifle firing pin (§2.2). The pin does not enter into any patterns involving distal features of the environment that are not perfectly matched by the proximal causal story. Movement of the trigger corresponds to movement of the pin corresponds to ignition of the primer, explosion of the propellant, and discharge of the bullet. Robust outcome functions 'bridge' to common outcomes across a range of different proximal conditions. That is absent in the case of the firing pin. (This is spelt out more carefully in §8.2b.)

Notice that standard teleosemantic accounts of content require a consequence etiology but do not require robust outcome functions. That misses out on an important element of the cluster that gives representations their explanatory bite. The honeybee nectar dance has evolutionary functions irrespective of any feedback that comes from collecting nectar. That qualifies as an (evolutionarily based) task function, but only if the distal outcomes (arriving at distant flowers and collecting nectar) are also robust outcomes. As a matter of fact, bees do rely on a variety of inputs before performing the waggle dance, and they do reach foraging locations robustly in the face of obstacles and variations in wind speed (Srinivasan et al. 1996). As far as I know there are also robust outcome functions in the other central examples relied on by Millikan. So, the cases do fall into a cluster that supports representational explanation. But Millikan's definition of function does not include a condition that functions should be outcomes that are robustly produced. To characterize the functions that underpin representational content in the honeybee nectar dance, and other cases of directly evolved animal signalling, we need to combine evolutionary functions with robust outcome functions for the same outcomes.[14]

3.7 Rival Accounts

Griffiths (2009) argues that analysing functions in terms of contribution to persistence delivers the wrong result in many cases (see also Artiga and Martinez 2016). Organisms have many phenotypes that are detrimental to their own survival and only make sense in terms of their contribution to fitness. Behaviour that promotes mating or feeds offspring at the expense of the individual's wellbeing are obvious examples. Griffiths's example is the heavy investment in a single mating season made by males of several species of small Australian marsupials, which greatly increases their risk of death (Bradley et al. 1980, Diamond 1982). An extreme example is the way some male spiders engage in mating despite the fact that they will end up being eaten by their female mate (Andrade 1996, Forster 1992).

No doubt there are lots of these cases in nature, with many involving representation: signalling between organisms (e.g. to achieve mating) or internal representations (e.g. of the conditions indicating that now is the time to pursue a mate). Contribution to persistence cannot help to underpin representational content in these cases. Our pluralist framework will cover such cases if the behaviour produced has evolved directly by natural selection. As with animal signalling discussed earlier, representation in these cases will be underpinned by a task function which conjoins robust outcome function with evolution-based stabilized function.

Griffiths makes a rival proposal. He has a forward-looking evolutionary approach. Functions are causal role functions that will contribute to reproduction of the organism (Griffiths 2009, p. 25). This is similar to Bigelow and Pargetter's earlier proposal

[14] Shea (2007b) argued that a similar move addresses the dormitive virtue problem with teleosemantic contents.

that functions are effects that give an organism a propensity to succeed under natural selection (Bigelow and Pargetter 1987).[15]

Unfortunately, the two objections made earlier to forward-looking accounts of contribution to persistence (§3.4d) are also decisive objections to forward-looking accounts of contribution to fitness. Whether an effect will contribute to fitness is heavily dependent on the context (of other organisms and the rest of the environment). Either evolutionary history comes back in to specify the relevant context (the one organisms of that type have evolved to deal with) or there are just too many effects that would contribute to fitness in some circumstance or other. Without relying on history, there is also considerable open-endedness about what should count as the system. This open-endedness is a good reason why accounts of evolutionary functions should be based on actual evolutionary history, not possible future or counterfactual contributions to fitness (Godfrey-Smith 1994b, Artiga 2014b). Nor is there any in-principle answer to the question of how we should count fitness prospectively (at the first generation, the second generation, or further).[16] A forward-looking approach also makes functions unsuited to figuring in a causal explanation of why an organism behaves as it does, as argued earlier in relation to forward-looking contributions to persistence. These considerations make forward-looking evolutionary functions inappropriate as a basis for representational content.

Griffiths's examples of behaviour that promotes fitness but is bad for persistence of the individual can be understood in terms of the (historically based) evolutionary function of the behaviour. This does mean that there will be cases where the two different approaches to function are pulling in different directions in the same organism. Representations involved in the spider's mate-approaching behaviour get their content in virtue of achieving a task function based on the way behaviour of that type has promoted offspring-production (hence fitness) in its ancestors. At the same time representations involved in the spider's homeostatic mechanisms can get their content from contribution to persistence, and also in virtue of having been reinforced by some basic learning mechanisms, both irrespective of their evolutionary functions (although in this case they are likely to have evolutionary functions as well). Intentional design can also produce task functions that conflict with evolutionary-based task functions. For example, by design we could use a glow worm as a light-sensitive switch to turn on the heating when it gets dark. So, our framework allows for task functions based on evolution which do not contribute to persistence (Griffiths's case), and also

[15] Nanay (2014) makes a related proposal: that the functions which teleosemantics should rely on can be analysed in terms of subjunctive conditionals about fitness: effects that would contribute to fitness of the organism.

[16] Standardly, fitness is measured in terms of expected long-run genetic contribution to the population, but whether that is the best measure for predicting evolutionary change over time will depend on the particular situation.

for task functions based on learning or contribution to persistence that have conferred no reproductive advantage.

3.8 Conclusion

In this chapter we have been examining one of the two key elements of the framework introduced in Chapter 2: the task being performed by a system. What counts as a system's tasks or functions—the functions whose performance is to be explained representationally? Answering that question is usefully constrained by the desideratum that an account of content should show why adverting to representational content allows better explanations of behaviour than would be available otherwise. Representation in many subpersonal systems forms part of a real cluster in nature, in which three elements are instantiated together, better than chance and for a natural reason. This cluster is what gives representational content its explanatory purchase. A central element in the cluster is a system's having a stabilized function: producing outcomes that have been stabilized by evolution, learning or contributing to the persistence of the organism that produces them. Stabilized functions tend also to be robust outcome functions, and the converse. The third element is that there is an internal mechanism which accounts for these outputs being stabilized and produced robustly, a mechanism in which internal components (representations) stand in exploitable relations to relevant features of the distal environment. In these cases we can see both how and why robust outcomes are successfully produced. The internal components are how, and the stabilization process is why. When the three elements are instantiated together, a sufficient condition for having representational content is met, and recognizing such contents affords better explanations of the behaviour of the system than would otherwise be available.

4

Correlational Information

4.1 Introduction

(a) Exploitable correlational information

Chapter 2 introduced a framework for understanding representational content. Chapter 3 filled in one half of the framework: the functions being performed by an organism or other system. The other half is having an internal organization that capitalizes on exploitable relations—relations between internal states and the world that are useful to the system. Not all task functions are achieved representationally. Representation arises where a system implements an algorithm for performing task functions. That in turn has two aspects: internal vehicles stand in exploitable relations to features of the environment which are relevant to performing the task; and processing occurs over those vehicles internally in a way that is appropriate given those relational properties. Content is constituted in part by exploitable relations: internal processing implements transitions between vehicles which make sense in the light of these relational properties, transitions called for by an algorithm which is suited to producing the input–output mapping given by the system's task functions. This chapter focuses on cases where correlation is the candidate exploitable relation. The next chapter looks at structural correspondence.[1]

The account shares with teleosemantics a reliance on teleofunctions (Chapter 3) and the insight that the way a representation is used downstream is important to fixing its content (for me, also the way it is produced). However, we will see that my account does not presuppose that there are dedicated representation consumers that play a special role in constituting content. That is an advantage of my view over some standard teleosemantic treatments (§1.4, §1.5).

An object or process carries correlational information just in case one or more of its properties correlates with the properties of some other object or process. More formally:

Correlational Information
Item(s)[2] a being in state F carries *correlational information* about b being in state G iff
$P(Gb|Fa) \neq P(Gb)$

When a carries correlational information, observing the state of a is potentially informative, at least to some extent, about the state of b. Such correlations are obviously useful, the more so the stronger they are; that is, the more a's state changes the probability of b's state.[3] An organism which needs to condition its behaviour on whether some state of the world obtains, but can't directly observe that state of the world, can instead condition its behaviour on the state of an item which carries correlational information about the relevant state of the world.

Our definition of correlational information relies on there being nomologically underpinned probabilities in the world (propensities, objective chances, nomologically based frequencies, or the like). An organism that observes a positive correlation between Fa and Gb can form an expectation, when next encountering an instance of Fa, that Gb is more likely. That expectation is well-founded if the reason for the correlation in the originally observed samples carries over to the new sample. It need not. Suppose that a particular shade of green that occurs on meat, *green*-123 say,

[1] We leave aside two other candidate exploitable relations because they don't arise in our simple systems: subject-predicate structure of the kind found in natural language (genuine singular and general terms); and the semantic or inferential connections between concepts, which potentially play a content-constituting role there. There may be more.

[2] I am deliberately neutral about what should count as an item. It could be a particular object, e.g. a = the flagpole on top of Buckingham Place and F = the Union Jack is flying. Or it could be a collection of objects or a type of object, e.g. a = human faces and F = having red spots. It could also be a process or a type of process.

[3] This 'change' need not be causal—'change' is simply convenient way of saying that the conditional probability differs from the unconditional probability.

is a sign of a bacterium which will multiply in the gut and lead to illness. A person could notice that eating something green-123 made them ill. They could well form the expectation that anything *green*-123 is poisonous and should not be eaten. As a result, they also avoid eating some vegetables. Suppose that, in plant leaves, green-123 happens to be a sign of a toxin produced by plants to discourage herbivores. Then green-123 in a leaf does indeed raise the probability that the leaf is poisonous to eat. But it is only by accident that the correlation that exists in meat extends to plants. An organism observing the correlation in meat and projecting an expectation to plants would get things right, but only by accident. There is no nomologically underpinned correlation which explains why the expectation formed in one case should carry over to the other.

We are interested in the correlations that can be exploited by an organism to learn from samples and project to new cases, so it should be non-accidental that correlations encountered in one region (in the history of the individual or its ancestors) should project to new cases. That point generalizes. We are going to rely on the way exploitable correlations figure in causal explanations of behaviour and its success. So, we need to have recourse to correlations where it is not an accident that the correlation extends from one region to another. The correlation must subsist for a univocal reason.

Correlations can be useful if they raise probability or if they lower probability, but not if they do so unpredictably. What is useful is if there is a region where probability is raised throughout that region, or if there is a region where probability is lowered throughout that region.

Accordingly, I define exploitable correlational information as follows:[4]

Exploitable Correlational Information
Item(s) a being in state F carries *exploitable correlational information* about b being in state G
iff
(i) there are regions D and D' such that if a is in D and b is in D',
$P(Gb|Fa) > P(Gb)$ for a univocal reason
or
(ii) there are regions D and D' such that if a is in D and b is in D',
$P(Gb|Fa) < P(Gb)$ for a univocal reason

'Region' is intended to connote a spatiotemporal region but can be understood more widely to include sets and other kinds of collection. Items a could be all members of a species, or even all organisms, or just one individual. Where a is a particular object, the region will just be the places and times where a is (or the singleton set whose only member is a). The region could be smaller. Anya-while-adolescent may exhibit a correlation between a facial expression and a subsequent behaviour. The relevant region would then be Anya during adolescence. The items a may also be a type of object, such as human facial expressions. The restriction to regions means

[4] Modifying Millikan's definition of 'soft natural information' (Millikan 2000, Appendix B).

there is no need for universality. The correlation may be highly local, such as facial expressions of Hoxton twentysomethings in the early 2010s. It is of course implicit that the items carrying exploitable correlational information are only the ones drawn from the relevant region.

The definition above is point-wise: one state raises the probability of another. In many natural cases a can be in a range of states, each of which raises the probability of b being in one of a range of other states. For example, the number of rings in a tree core correlates with the age of the tree: there being two rings makes it probable that the tree is two years old; three rings, three years old; and so on. Then F and G above can take a range of values, with each value of F mapping to a corresponding value of G about which it raises the probability.[5] An organism may learn or evolve to make use of this systematic relationship. It can then extend that expectation to new instances of the same overall relationship. A person could observe a few instances of the correlation between tree rings and age and then form the general expectation that tree age is equal to the number of rings. They may never have encountered forty-two rings in a tree core before; nevertheless, when they count forty-two rings and form the expectation that the tree is forty-two years old, that expectation is correct for an underlying univocal reason that extends from the cases they learnt about to the new case.[6] A further feature is that the different states *X* that a may be in exclude one another: any particular a can only be in one of these states at a time. In many cases they form a partition, covering all the possibilities (e.g. all possible numbers of tree rings). We can define a notion of exploitable correlational information carried by a range of states as follows:

> *Exploitable Correlational Information Carried by a Range of States*
> Item(s) a being in states *X* carries *exploitable correlational information* about b being in states *Y*
> iff
> there are regions D and D' such that, if a is in D and b is in D', for a univocal reason, for every value F of *X* there is some value G of *Y* such that P(Gb|Fa) > P(Gb) or P(Gb|Fa) < P(Gb)[7]

[5] This is closely related to Shannon's theory of information, which connects a range of states of a receiver with a range of states of a source. Our point-wise correlational information is a special case of this. Shannon information additionally takes account of the probability distribution across this range of states.

[6] Millikan (1984) made this important observation about evolutionary functions. Suppose that a bee dance of exactly 42.5° to the vertical has never been performed in the history of the honeybee. Nevertheless, this dance has the evolutionary function of sending consumer bees off at 42.5° to the direction of the sun. The function of the particular dance derives from the systematic relationship bees evolved to respond to (angle of dance to the vertical = direction of nectar in relation to the sun). I am making a parallel point about exploitable correlations. Where there is a univocally grounded, nomologically underpinned, systematic probability-raising relationship, expectations acquired for some values can be extended, non-accidentally, to other values drawn from the same system.

[7] For a particular pair F and G, Fa must raise the probability of Gb across the region or lower it across the region; raising probability in some subregions and lowering it in others would not count. (This is explicit in the definition of exploitable correlational information above.) But the mapping from values of

Animal signalling is an obvious case where correlations are exploited in the service of a function—in those cases, an evolutionary function. If there is also robustness in how the outcomes prompted by the signals are achieved, which there often is, then these cases will fit squarely within our framework. As we saw in discussing teleosemantics (§1.4), the correlations that feed into an explanation of how behaviour prompted by signals achieves its evolutionary function are correlations with distal features of the environment (e.g. with the location of nectar). Skyrms-style signalling models also turn on correlations being exploited as stand-ins on which receivers can condition behaviour (Skyrms 2010, Shea et al. 2017). (These models abstract away from the machinery of robustness.) There, which correlations are relevant can be read off directly from the payoff matrix: correlations with world states in which, given appropriate actions, payoffs are delivered.

The definition of exploitable correlational information is extremely liberal. There are very many different regions within which a correlation subsists. There will often be subregions where a correlation is stronger and larger regions where it is weaker; also partially overlapping regions. I don't attempt to define a unique reference class with respect to which a univocally based correlation exists. There is exploitable correlational information with respect to any region within which a univocal reason extends, whether or not that reason also extends to a wider region. What counts is the region within which an organism operates: the instances of a it encounters and the instances of b on which success of its behaviour depends. The basis for the correlation is objective and independent of the organism, but the correlation strength that matters partly depends on the organism's point of view.

For our purposes below, there has to be an exploitable correlation within the region where outputs were stabilized and robustly produced. And the correlation encountered there has to be strong enough to explain stabilization and/or robustness. What counts as strong enough will depend on the facts of the case.[8] An extremely weak correlation might form an adaptive basis for avoiding predators, for example, because the costs of being eaten are so high. What matters in explaining stabilization is that the correlation is strong enough in the region in which stabilization occurs. If we are instead looking forward, predicting the likelihood that its behaviour will be successful, then it is correlation strength in the region where the organism will be operating that is relevant.

Correlation has been the focus of a lot of scientific work on representation in the brain.[9] At the level of individual neurons, neuroscientists have consistently looked

X to values of Y may be such as to raise the probability for some values of X and Y and to lower it for other values of X and Y.

[8] There are also several aspects of correlation strength that will be important in different ways: sensitivity, specificity, positive predictive value, negative predictive value, etc. The following are always important: how likely the world state Gb is given the vehicle state Fa, i.e. P(Gb|Fa), and how informative the vehicle state is about the world state, i.e. how different P(*Gb*|*Fa*) is from the unconditional probability P(*Gb*).

[9] In 'information processing' or 'computational' theories in psychology and cognitive science, the 'information' is usually a matter of representational content rather than bare correlation.

for correlations between neural firing rate and certain kinds of stimuli; for example, neurons that respond to an edge in a specific location in the visual field (Hubel and Wiesel 1962).[10] In standard region-based fMRI the search has been for regions of the brain whose activity correlates with a particular type of stimulus or task. More recent multi-voxel pattern analysis looks for correlations between the distributed pattern of activation across a region of interest and a stimulus or task type. And model-based fMRI looks for regions whose activation varies parametrically in step with quantities that the brain may be computing.

All these techniques are probing the way neural activity carries correlational information. Three features of these practices are worth noting. First, strength of correlation is always assumed to be important: carrying more information is, *ceteris paribus*, more useful; and so it is assumed that strong correlational information is a better candidate for what the brain is really representing. Secondly, the correlations being probed are very often with distal features of the environment: properties of the stimuli being presented or the task the organism is called on to perform. Thirdly, there often seems to be a tacit assumption that only information that is being used is relevant to understanding what the brain is computing (deCharms and Zador 2000). For example, there might be substantial information carried by the phase difference between neural firing rates, but that is of no interest unless there is a way for downstream neurons to detect and make use of those phase differences. That is at the level of the vehicle, and a similar constraint is often in the background at the level of the content. Incidental correlational information that just happens to be carried by a pattern of neural firing is not a candidate to figure in the computational or information-processing story unless it is somehow relevant to how the organism is behaving (Hunt et al. 2012).

(b) Toy example

Before putting forward a concrete proposal about how correlational information gives rise to content, let's look at a simple example in which correlation is being exploited by a system to perform a task. Consider the toy system from the last chapter which moves along a line until it reaches a point *T*, where it stops (§3.6a).

Our toy system has four internal vehicles, **t**, **r**, **δ** and **a** (Figure 4.1). In the final version, **t** initially varies randomly across multiple episodes of behaviour, until its value is fixed by a recharge. Because of this, the value which **t** eventually adopts correlates with the location of a power source. The obvious useful correlations then are those given in Table 4.1:

[10] One type of correlation is where maximum firing rate corresponds to a particular feature at a particular location, dropping off with distance or variation in the feature (e.g. rotation of a line). Another type of correlation is filtering, where it is not the maximum firing rate that is most important, but the sensitivity of changes in firing rate to changes in the stimulus. A neuron whose firing rate goes up and down substantially as the orientation of a bar changes, say, will thereby carry fine-grained information about orientation. The orientation to which it is most sensitive will be somewhere in the middle of its range of firing rates, not at the maximum.

Table 4.1. Useful correlations carried by components of the toy system

Vehicle	Correlation
r	system's position on the line
t	location of a power source on the line
δ	distance of the system from a power source
a	velocity with which the system moves along the line

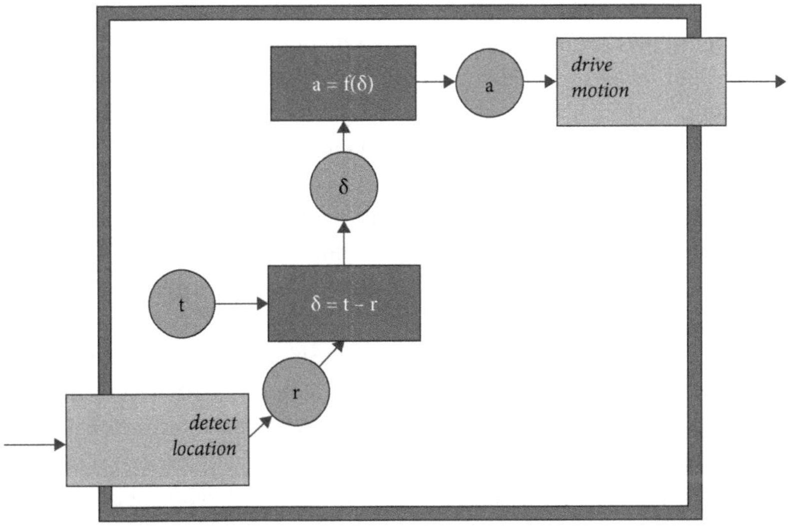

Figure 4.1 Toy system discussed in the text.

Those correlations make the performance of the system intelligible. That is the heart of the reason why the vehicles listed on the left of Table 4.1 are representations and the conditions on the right are their respective contents. Firing rate (let us say) of vehicle **r** correlates with distance from the origin, of vehicle **t** with location of a power source. Therefore, vehicle **δ**, whose firing rate correlates with the difference between the firing rates of **r** and **t**, will correlate with the distance of the system from a power source. The firing rate of vehicle **a** is proportional to that distance. If that rate is linearly transformed in an appropriate way into velocity, then the system will move from any point along the line so as to reach the power source. Given that those four internal elements carry the correlational information listed above, internal processing over those elements, which proceeds in virtue of vehicle properties, constitutes an algorithm for performing the distally characterized task being performed by the system (reaching T). These contents meet our desideratum (§2.2): they allow us to see why representational contents enable a better explanation of the system's behaviour than would be available without them.

Those internal components carry lots of other correlational information—information that is less relevant to explaining how the system performs the task. For example, **r** correlates with the activity of some sensory receptors just upstream on the system's periphery. That correlation would also help explain performance, but only when supplemented with the fact that activity of the sensory receptors correlates with position along the line. So the correlation between **r** and sensory stimulation would figure in a less direct explanation of how the system performs the task. On the output side, **a** correlates with the speed of rotation of the wheels. That correlation is only explanatorily relevant because rotation of the wheels correlates with the velocity at which the system moves. So that too is less directly explanatory of how the internal components conspire together to allow the system to achieve its task function.

Suppose light falls on the engineer's workbench from one side, diminishing in intensity along the bench. Then component **r** correlates with the intensity of light at the toy's location and component **t** with light intensity at the power source. These distal correlations would also explain why the difference **δ** correlates with the distance of the system to a power source, but only when supplemented with the information that light intensity correlates with distance along the bench. So, this set of correlations carried by the internal components offers, collectively, a less direct explanation of how the system performs its task function.

We should beware of seeking determinacy where it's the wrong place to find it. And, indeed, there is some indeterminacy in what this simple toy system represents, according to my account. There are other collections of correlational information that are just as good for explaining task performance—that collectively figure in just as direct an explanation: between **t** and the location of *something worth reaching* (and between **δ** and the distance to *something worth reaching*) for example; or between **t** and the location of *a place where an outcome that reinforces behaviour and promotes persistence occurs* (with a corresponding correlation at **δ**). These alternative contents are not equivalent, since they could come apart, but they make distinctions that are more fine-grained than those that are relevant to the system. In the life of this very simple toy system being a place worth reaching is coextensional with being a charging point. Component **t** is correlated equally strongly with both. As theorists we should say that the content represented by the system is indeterminate between these options.[11] We will deal with indeterminacy in more detail in §6.2, with the benefit of the positive accounts of content set out in this chapter and the next.

[11] There are different ways of capturing that indeterminacy. One is to say the representation has a correctness condition which is the disjunction of these conditions. An alternative is to say that the words we theorists use to describe the correctness condition are only an imperfect expression or model of the true correctness condition; and that each way of capturing the correctness condition using the fine-grained machinery of natural language is bound to be only approximate, each equally good.

4.2 Unmediated Explanatory Information

(a) Explaining task functions

In looking at a toy example in the last section, we saw that not all correlations are on a par for the purposes of explaining how a system manages to achieve its task functions. Some correlational information carried by internal components figures directly in an explanation of how a system with such internal processing is able to perform the task and become stabilized by feedback; other correlational information is only more indirectly explanatory; and some is explanatorily irrelevant. Recall that the underlying motivation for representationalism—the practice of adverting to content properties carried by real internal components to explain behaviour—is the idea that the system's internal organization implements an algorithm for performing a task being carried out by the system. Correlations between internal elements and distal features of the environment show how a system's internal organization is keyed into the world so as to perform the distally characterized task. Content fixed in that way would meet our desideratum (to produce a theory which allows us to see how contents explain behaviour). So, the correlations that are content-constituting should be those which explain how the system achieves task functions (i.e. stabilization and robustness).

The move I make here involves a subtle shift of perspective. One could hold that content is fixed directly by its role in representational explanation: a system represents whichever contents best account for the pattern of behaviour the system produces.[12] Rather than representational explanation, my account is founded on causal explanation. Which correlations figure in causal explanations of stabilization and robustness? The former approach makes a very tight connection between what contents are and what contents explain, generating considerable indeterminacy. Causal explanations of stabilization and robustness are less indeterminate (§4.1a, §6.2).

To put this more carefully, we first define the explanandum, using a term of art, 'explaining S's performance of task functions'; then we define 'unmediated explanatory information', which is the correlational information that figures in the explanans. The explanandum has two elements, corresponding to the two elements of task function (§3.5). First, we can explain how outcomes have been stabilized (hence count as stabilized functions). Secondly, we can explain how outcomes are robustly produced (hence count as robust outcome functions). No single term is perfectly suited to encompass these two explananda. In a sense we are explaining why an outcome F is a task function of a system S, but that in turn calls for an explanation of how it was stabilized and robustly produced, so 'why' becomes somewhat misleading. 'Explaining performance of task functions' is neutral enough to cover both explananda. It also

[12] That would make it hard to meet our desideratum.

emphasizes that we are focused on explaining how the system does or has done something (in its environment).

Explanandum
To explain *S's performance of task functions F_j* is to explain:
(a) how producing each of the F_j has been systematically stabilized through evolution,[13] learning or contribution to persistence (see §3.4d);

and/or
(b) how each F_j has been produced in response to a range of different inputs and achieved in a range of different relevant external conditions

Unmediated Explanatory Information
The *UE information* carried by a set of components R_i in a system S with task functions F_j
is
the exploitable correlational information carried by the R_i which plays an unmediated role in explaining, through the R_i implementing an algorithm, S's performance of task functions F_j

The idea that some correlations play an unmediated role in an explanation calls for clarification. In the classic example of the frog's fly-catching mechanism, the correlation of retinal ganglion cell firing (R) with little black things figures in *an* explanation of how the system was stabilized by evolution, but that explanation also mentions the fact that being a little black thing (condition C) correlates with being a nutritious flying object (condition C'). Without that background correlation, it would be opaque how the correlation between R and C enabled frogs to achieve an evolutionarily beneficial outcome. So, the role of the R–C correlation in that explanation is mediated. There is another explanation of stabilization that adverts directly to the correlation between R and nutritious flying objects (C'). The role of the R–C' correlation in that explanation is unmediated. A correlation between an item R and condition C plays a *mediated* role in an explanation if its role depends on the explanation adverting to a further correlation between C and some further condition C'; otherwise it plays an *unmediated* role.

The discussion in the last section effectively argued that the correlations set out in Table 1 qualify as UE information carried by the components of our toy system. (It also argued that this list is not exhaustive: there are other sets of UE information carried by the same components, hence some indeterminacy.) My claim is that, where correlation is the relevant exploitable relation, the correlational information that is content-constituting is UE information. More specifically, a sufficient condition for a vehicle to represent content p is that it carries UE information about p.

[13] Recall that in this case 'S' picks out a lineage of organisms, typed by a lineage-based category (§3.3). Then we have explananda like *how honeybees reach locations of nectar*, or *how* E. coli *bacteria avoid toxic chemicals*. Species (e.g. honeybee, *E. coli*) are lineage-based classes of organisms.

Condition for Content based on Correlational Information
If component R of a system S with task function or functions F_j carries UE information about condition C,
then R represents C

There is no need for an account of content to accord with what scientists relying on content think it is. For example, scientists may have no idea that content is connected to functions which are partly historically based. Nevertheless it is interesting to see that my theory of how content is constituted closely parallels a recently developed method for finding out which computations are being performed in a neural system, the method of model-based fMRI (Corrado et al. 2009). The method starts with behavioural data. For example, subjects may be asked to choose between pairs of fractal images, where different images are more or less likely to be rewarded. Subjects learn through feedback which images are rewarded when. The probabilities change during the experiment and subjects' behaviour adjusts accordingly. A large number of choices produces a rich source of data about how subjects' choices are influenced by the history of feedback they have received for past choices.

The first step is to find which computations the subjects could be performing: algorithms that are capable of producing the observed pattern of behaviour. In our terms, that is to find a list of candidate algorithms that could perform the task functions these organisms have been trained to perform. The second step is to go into the brain to see which potential algorithm is most consistent with neural activity. An algorithm calls for various quantities to be computed on the way to making a choice: expected reward, reward received at this time step, prediction error, adaptive learning rate, etc. The fMRI BOLD signal reflects the amount of neural activity in small areas of the brain, hence can reflect the quantities being represented by an algorithm implemented in the brain. We look to see if there are areas of the brain whose activity varies, trial-by-trial, with the varying quantities called for by a candidate algorithm. When areas show up as potentially representing quantities called for by the algorithm, we check that it is plausible, in terms of neural architecture, that they are computing those quantities in the right sequence. This process is repeated for other candidate algorithms and then a 'model comparison' is performed to see which algorithm is most consistent with the neural data. There are many assumptions behind the method, not all of which are strongly supported yet, but nevertheless when algorithm A fits the behavioural and neural data better than rivals B and C, that gives us some reasonable evidence that the brain is implementing algorithm A rather than B or C (Mars et al. 2012). What is striking for our purposes is that the method is effectively looking for correlational information in the brain which explains how a person performs the task function observed in their behaviour. Model-based fMRI is looking for the properties which, according to varitel semantics, are constitutive of content.

An implementations of an algorithm has a dual character, one aspect purely local to the system and another that depends on relational properties of components of the system. So, having components carrying relevant correlational information is only one half of what it takes to implement an algorithm. The other is that they should be processed in the right way—in a way that makes sense in the light of the correlational information they carry and will thereby generate appropriate behaviour. That is, when the processing is characterized in terms of local properties, independent of correlational information carried by the vehicles, it should proceed through the steps called for by the algorithm.

That in turn puts tight constraints on which correlations are likely to be explanatory, since an algorithm usually calls for different vehicles to be doing different things. For example, an algorithm might call for one vehicle that correlates with shape and another with colour, putting that information together in a third vehicle that correlates with object category. There would be *an* assignment of content to vehicles according to which all three steps simply register object category, quite noisily in the first two cases. But that set of correlational information carried by components implements a less explanatory algorithm—an algorithm that does a less good job of explaining how the system performs its task functions. For this reason, an algorithm that relies on different vehicles carrying different correlational information will generally be more suited to explaining task performance (§6.2f). We saw that at work in our toy example: an explanatory set of correlational information has **r** correlating with position and **δ** with distance to the power source (rather than both registering distance to the power source, say). In the definition of UE information, 'unmediated' does not count against this. To count as explanatory, an algorithm will generally have different contents at different stages. The computation of what to do is mediated through a complex series of internal states, but the job of each should be to keep track of an external condition directly, not in a way that depends on presuming a further background correlation holds.

UE information covers output correlations as well as those that are due to how a system responds to inputs. The algorithm in our toy system relies on the fact that **a** drives the toy with a certain velocity: it correlates with velocity by causing motion. Not all UE information can be about outputs, however. Part of explaining performance of a task function is to explain robustness, in particular how an output was produced in response to a range of different inputs. That will require some components of the system to carry correlational information that they reflect rather than produce. To anticipate the distinction we will discuss in Chapter 7, components can have directive contents when UE information concerns outputs, but there has to be some descriptive content in the system somewhere.

Often output-based UE information will concern the means by which a task function is performed. In our toy example, moving with a certain velocity is a means by which the toy reaches a power source from a range of different starting positions. However, sometimes the relevant correlation will be to output an F which is itself a task function of the system. For example, humans have a learning-based task function of getting sugar (in circumstances when it is needed and available). In calculating how to do that, it looks like we have an internal state in orbitofrontal cortex whose job is to correlate at input with whether we need sugar and to correlate at output with obtaining sugar (Rolls 2015, Rushworth et al. 2011, Alexander and Brown 2011). How can that output correlation be explanatory of the task function? Isn't it identical with the task function? The answer is that UE information falls out of the way the whole internal mechanism explains how outputs are produced robustly and stabilized. That requires more than just producing output F. It requires producing F in a range of different circumstances and keying it into the environmental circumstances in which it was stabilized. The algorithm as a whole is explanatory of that. A component correlating with F is one part of the overall explanation, but only when combined with other components carrying other UE information, some of which will have descriptive content. (Recall again, we are not asking which contents would best explain the behaviour; UE information is based on how an internal mechanism forms part of a causal explanation of robustness and stabilization.)[14]

This account of the way correlation can ground content is very much in the spirit of Dretske (1986, 1988). Dretske considers the case where an internal component correlates with a feature about which the system is disposed to learn, in the sense that instrumental conditioning will shape the system so as to condition its behaviour on that feature. For example, it could be because an internal state correlates with the location of a peanut (on the left or the right) that the animal comes to condition its reaching behaviour on that internal state (reaching left or right, respectively). Dretske calls the correlation with the location of the peanut a 'structuring cause' of the system's later behaviour.

That is one version of the idea that explanatory connections between correlations and the stabilization of behaviour are relevant to content determination. However, I have a more general account of why Dretske's recipe produces the right answer in the case he deals with. It is because of the role of correlations in explaining stabilization and the establishment of task functions. Instrumental conditioning of the kind Dretske points to is one specific example of that. My account is more general in three respects. First, it applies to a wider range of cases than just those which involve instrumental conditioning.[15] Secondly, my view does not require there to be pre-existing correlations

[14] A different kind of case is more straightforward to deal with. Sending consumer bees off foraging 200 m away in the direction of the sun when there is nectar at that location is a task function of a bee colony. It is an output that was stabilized by evolution and robustly produced. A dance of four waggles (say) in a vertical direction correlates with sending consumer bees foraging at that location. That output correlation is with an F which is a task function. But it also explains how the colony achieves another, more general, task function: getting nectar (from a variety of different locations). So, some correlations with task-functional outputs get explanatory purchase through explaining other, related task functions.

[15] Dretske does allow that natural selection can give an internal state the function of indicating something. The internal state can then be called a representation. But he argues, mistakenly in my view, that this is not a case where contents (reasons in his terminology) explain behaviour, since what the states indicate, 'is (and was) irrelevant to what movements they produce' (1988, p. 94), see also Dretske (1991, pp. 206–7).

between internal states and distal features. The correlations could develop at the same time as the system is being tuned to behave in a certain way. That is what happens when an artificial neural network is trained, for example. Thirdly, it applies to cases where several different correlational vehicles are involved in generating behaviour, as in the toy example we have been discussing. Dretske's recipe only applies straightforwardly where one correlational vehicle comes to be wired up to drive behaviour in virtue of the correlational information it carries.

The latter point is important, because any plausible account of representation in the brain will have to deal with the fact that very many representational vehicles interact in complex ways to produce behaviour. In §4.4 we will see a variety of ways that can arise.

(b) Reliance on explanation

The definition of unmediated explanatory information (UE information) places heavy reliance on the concept of explanation (obviously). It bases content on causal-explanatory connections. I am assuming a realist account of explanation according to which the causal-explanatory relations that figure in explanations are objective metaphysical dependence relations.[16, 17] This is not special pleading. Varitel semantics is making use of a resource here which other sciences also take for granted. It is not the task of a theory of content to give a theory of why causal-explanatory relations are objective.

Recall that contents are fixed, not by the role of contents in representational explanations, but by the role of correlations in causal explanations. So, my theory of content is not interest-relative or pragmatic. If the definition of UE information is not empty, then it picks out a property in the world. UE information then exists, irrespective of whether anyone chooses to refer to it. It might be an interest-relative matter whether we go in for explanations that appeal to this property, that is, whether we go in for representational explanations. I have been arguing that UE information (and UE structural correspondence, in the next chapter) underpins a distinctive scheme of explanation, one where correctness explains success and misrepresentation explains failure. Our epistemic interests may affect whether we appeal to this scheme of explanation.

If I'm wrong to assume that causal-explanatory relations are objective, then my accounts of content necessarily inherit any interest-relativity of causal explanation. However, the same would then be true throughout the sciences. If causal-explanatory

[16] This is compatible with the view that explanations are semantic entities (e.g. sentences, models); as well as with the 'ontic' view of explanation (Salmon 1984, Craver 2014).

[17] In causal explanations, 'explains' does not introduce an intensional context, in the sense that in an intensional context it matters how we pick out the properties referred to. A neuron in a macaque's OFC that carries UE information about *orange juice* thereby carries UE information about *my favourite juice* (as it happens). Explanation does not of course in general allow substitution *salva veritate* of one property for another property which has the same extension. (In that sense causal explanations do not in general allow substitution *salva veritate* of coextensional property terms.)

claims in all sciences are ineliminably interest-relative, then it would be no surprise that representational contents are no different.[18]

Defining a property in the way I have always raises another pressing question. It's not enough to just show that the property exists (the definition is non-empty), and exists independently of anyone's interests. Is the definition any use, does it pick out a worthwhile category? I say 'yes', of course. My argument is that UE information meets our desideratum. It allows us to explain how representational content can explain behaviour. UE information is thus a worthwhile property because, if I'm right, it is the property that figures in many explanations in cognitive science.

(c) *Evidential test*

Carrying UE information not only explains, but also makes it more likely that the system will achieve its task functions. That gives us another way of getting at the UE information carried by a vehicle: increasing the correlation strength of UE information should increase the likelihood of the system achieving its task functions; similarly, weakening the correlation should decrease it. So, we can supplement the constitutive condition above with an evidential test—a (fallible) way of working out what UE information is carried by elements of a system.

> *Evidence of UE information*
> For component R in a system S performing a task function or functions F_j
> the correlation of the state of R with a condition G involving natural properties and objects in S's environment
> whose strengthening most increases and whose weakening most decreases the likelihood of S achieving F_j
> is a good candidate to be UE information carried by R

To see how this works, let's go back to our toy system. Suppose there is some random noise in the system, so that each component has a small chance of going into a random state during an episode of behaviour. Then the probability that the system is at location x, say, when **r** is in a particular state R_1, although high, is not certain. There will be some occasions when **r** is in R_1 and the system is in fact at random other locations. On those occasions, the system would not achieve its task function of reaching *T*. If the correlation between **r**'s being in state R_1 and the system's being at location x were strengthened, the system would achieve its task function more often. Similarly, weakening that correlation (increasing the noise) would reduce the frequency with which the system would reach *T*.

Now consider the correlation between **r** and light intensity. Strengthening that correlation might increase the likelihood of the system reaching *T*, provided the light intensity gradient is reasonably stable, but not by as much as strengthening the correlation with the system's location on the line would (since light intensity is not a

[18] As we've just seen, the relevant interests would be those related to giving causal explanations (of stabilization and robustness), rather than interests related to giving content-based explanations.

perfect correlate of location). So, the evidential test suggests that the correlation with light intensity is a less good candidate to be UE information.

Correlations on the output side can also be assessed using the evidential test. Where slippage in the wheels or noise in the motor system impairs performance, strengthening the correlation between component **a** and the velocity of the system will have the biggest effect in increasing the likelihood that the system performs its task function of reaching *T*.

The evidential test uses the effect of correlation strength on likelihood of achieving task functions as a proxy for how directly a collection of correlations carried by components explains the achievement of those functions. However, those things are not bound to align. Nor is there a guarantee that there will be an item of correlational information that satisfies the evidential test. A correlation whose strengthening improves performance may not be such that its weakening reduces performance, for example if there is a backup mechanism that puts an effective floor on the likelihood of performing a function. Even when there is correlational information that satisfies the evidential test, that is no guarantee that carrying this information figures in an unmediated explanation of performing task functions.[19]

The evidential test is restricted to correlations with natural properties, in order to focus on correlations that are candidates to figure in a causal explanation of task performance. General principles about explanation will make complex disjunctive or gruesome properties poor candidates to figure in such explanations. (Other theorists of content have also pointed to such considerations as ruling out some problematic putative contents.) There will clearly be correlations with non-natural properties whose strengthening would do more to increase the likelihood of success. In our toy example, if the state of vehicle **r** correlated with the location of the system *and* there being no noise anywhere in the system, then success would become very much more likely. These kinds of constructed properties are much less good candidates to figure in a causal explanation of stabilization and robustness, and hence less good candidates for content.

To apply the evidential test, we need first to have a collection of candidate correlations that assigns different correlations to different vehicles. As we've just seen, that is needed if implementing the algorithm (internal processing over components) is going to explain how outputs were produced robustly and stabilized by interactions with the environment. Then we fix on one candidate correlation, hold everything else fixed, and consider what would happen if the world were changed to make that correlation stronger. Rather than being at location x 95 per cent of the time when component **r** is in state R_1, what would the effect on robustness and stabilization be if it were at location x 100 per cent of the time when **r** is in R_1? In that case, the task-functional output (getting to the power source) would have been produced more robustly and would still have

[19] Suppose a computational step depends on comparing the values of two noisy representations and selecting the larger (as in analogue magnitude comparisons, see §4.6a below). If the noise is asymmetric around the mean, then reducing the noise in one register might cause the system to select the wrong option more often.

been stabilized. Strengthening the correlation of **r** with patterns of sensory input would have less of an effect on successful task performance, since sensory input is itself an imperfect correlate of location. So, the evidential test suggests that **r** carries UE information about the toy's location.

It is most appropriate to assess these counterfactuals against the circumstances that obtained during episodes of stabilization. However, it is only an epistemic test. Assessing what would happen to the system in its current circumstances will also give us some evidence, to the extent that the system's current environment is relevantly similar to the circumstances in which its behaviour was stabilized.

4.3 Feedforward Hierarchical Processing

In the next six sections we see how this account of UE information can be applied to a variety of case studies. The first case is where there is simply feedforward hierarchical processing of sensory input through a series of layers. Marr's account of 3D vision is a well-known example: inputs are processed into an array of point-intensities, then into a 'primal sketch' involving detectors for blobs, edges, and so on, then on into detectors for local surfaces and their orientations, and so on (Marr 1982). Hierarchical structure is also found in the successive layers of the artificial neural networks that have used 'deep convolutional' learning algorithms so effectively to categorize a huge array of natural visual scenes (Krizhevsky et al. 2012, Kriegeskorte 2015). To work with a simpler case, consider the ALCOVE neural network model (Kruschke 1992; see Figure 4.2).

The task of ALCOVE is to categorize objects, using its sensitivity to perceptual features. To give it a clear task function, let us suppose that it has been trained to sort objects into boxes according to whether the object falls under category A or category B. The training regime gives rise to task functions because internal configurations of connection weights that tend to produce incorrect behaviour are replaced, and those which produce correct behaviour are stabilized. As a result of training, the system can use its input-sensitivity to brightness, size, etc. in order to sort objects into the correct box. Putting an object of category A into box A is then a task function of the trained system.

It performs that function by taking an intermediate step before performing the sorting action. Training produces an array of 'exemplar nodes' at the network's hidden layer. These act a bit like names for individual objects. Activation of each correlates with encountering a particular object. The network solves the problem by first recognizing which individual object it is faced with, then sorting that object into the appropriate category. Input nodes correlate with features of the objects. Output nodes correlate with whether the object falls under category A or category B; they also correlate with where the object gets placed. Consider the correlational information carried by one of the exemplar nodes. Its activation raises the probability that:

(i) input nodes are activated thus-and-so
(ii) the object encountered has visual features abc (those characteristic of exemplar 1)

Figure 4.2 The ALCOVE network (Krushke 1992).

(iii) the object encountered is exemplar 1
(iv) the object encountered has visual features xyz (those characteristic of objects in category A)
(v) the object encountered belongs to category A
(vi) the object encountered will be placed in box A

Those correlations are presented in decreasing order of strength ((v) and (vi) are equal). Consider how these can be combined with the correlations carried at input and output so as to implement an algorithm for sorting objects into boxes. Correlation (iii) fits together with the input correlation with object features and the output correlation with object category to instantiate an algorithm for performing the task. The correlation of hidden layer nodes with groups of perceptual features (ii) would make for a less explanatory set of correlations. An algorithm that correlated with object category (v) at the hidden layer and then again at the output layer would be less explanatory of how the system actually manages to compute what to do and perform robustly. So (iii) is UE information carried by the hidden layer. These considerations also imply that the output layer carries the UE information given by clauses (v) and (vi) above.

How does the evidential test apply to this case? At the output layer, it is indeed the correlation with category whose strengthening and weakening most strongly affects the likelihood of success (excluding non-natural properties that would have an even tighter connection). At the input layer, it is reducing noise in the correlations with perceptual features that would have the strongest effect. Making an input node correlate more closely with exemplar or category would help in some circumstances but hinder in others, since input nodes are activated by more than one exemplar and more than one category.

The evidential test is equivocal when applied to the hidden layer. Because there is a straightforward many-to-one mapping from exemplars to categories, mistakes at the hidden layer that confuse one exemplar for another in the same category do not compromise overall performance of the system. So, tightening the connection between a hidden layer node and exemplar (iii) or category (v) will both improve performance, and to the same extent. To decide between them we have to turn to the consideration just mentioned: one collection of correlations (perceptual features, exemplar, category) provides a better understanding of how the system performs its task function than the other (perceptual features, category, category), since the latter overlooks some of the internal structure used by the system to perform the task (see also §4.1a).

Basing content on UE information has two general effects in these cases. It can select amongst coextensive properties about which the system carries information so that the most explanatory one figures in the content. And, since it is connected with explaining distal results achieved by the system, it tends to deliver contents that concern distal properties—properties that are relevant to how the system performs its tasks. So if we take JIM, a more sophisticated development of ALCOVE with more layers of processing, there is a layer of processing that detects geons—certain configurations of 3D shapes that objects can exemplify (Hummel and Biederman 1992). Does this layer represent properties of objects, or does it instead represent regular ways in which objects affect the system's sensory apparatus? If content is fixed by UE information carried by the layer, then it will be representing the former: properties of the distal objects.

A further development of ALCOVE uses a network with feedback connections between layers (Love et al. 2004). This raises the problem of reciprocal processing connections, which we will turn to with a different case study in §4.8 below.

It is worth noting that we have given an account of content for this system without having to give representation consumers a content-constituting role. Content comes out of how all the components interact to achieve task functions. In standard teleosemantics a consumer system is a special component, the evolutionary functions of whose outputs determine content. We noted above that it is not obvious how to extend the consumer idea to more complex cases (§1.5). That problem is not acute in this first, straightforward case study, but by eschewing a content-determining consumer even here we have an approach which is readily extendable to more complex cases.

4.4 Taxonomy of Cases

In §§ 4.5–4.8 we see how the UE approach can be applied to various cases from the empirical literature. A quick look at a typical wiring diagram for even a simple neural processing system shows that representation processing in the brain takes place in complex ways. The diagram of the primate visual system below is representative (Figure 4.3).

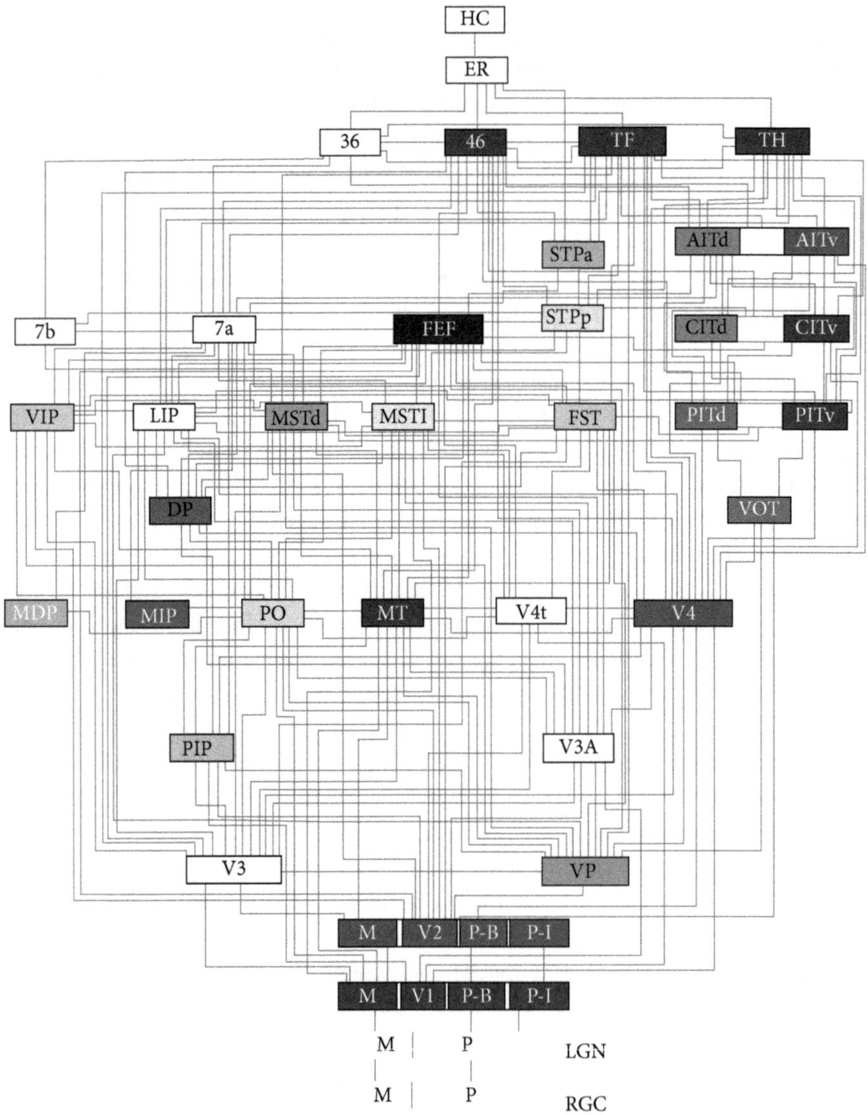

Figure 4.3 Diagram of the primate visual system (from Felleman and van Essen 1991).

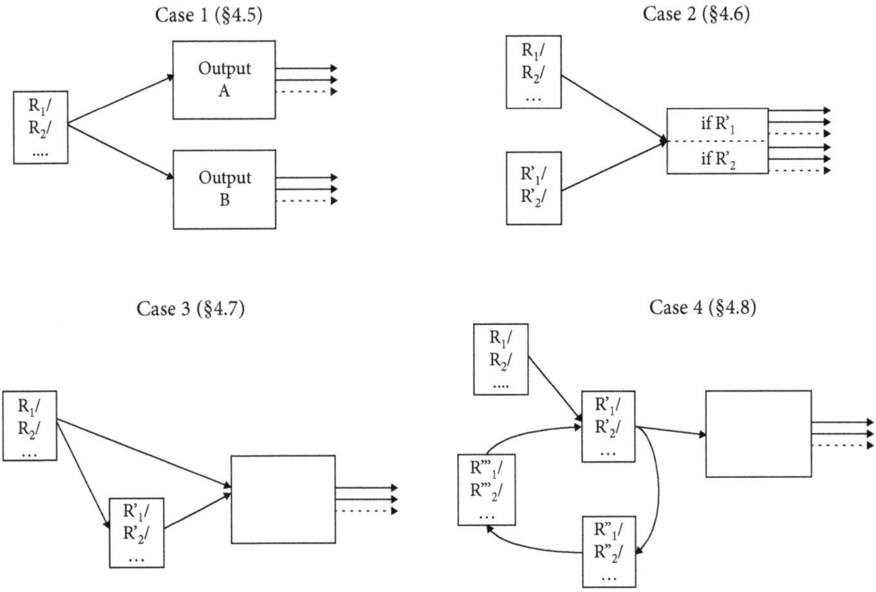

Figure 4.4 The four kinds of cases exemplified in §§ 4.5–4.8, respectively. (Inputs to the representational vehicles R and R' are not shown.)

I pick out four kinds of complexity that a theory of content will have to deal with—ways that representations are processed that show up regularly in information processing theories in cognitive neuroscience (Figure 4.4). Sections 4.5–4.8 then select case studies that exemplify each structure and show that the UE approach delivers appropriate contents. The cases are not exhaustive, but they do serve to demonstrate that the approach can be applied across a broad range of systems.

These cases also serve to highlight an important contrast with standard, consumer-based teleosemantics. We saw in the previous section that teleosemantics already faces a problem when we move away from cases like animal signalling where a single representational stage mediates between producer and consumer. When there are multiple layers of processing it faces a challenge in explaining why different stages represent different aspects of the problem. The consumer-based approach has an even bigger problem dealing with the kinds of cases taxonomized here, which are merely simplified treatments of some of the complex interconnections found in a typical neural processing diagram. The absence of a simple hierarchy and the presence of feedback loops make it very hard to identify, for each putative representation, a single consumer system which conditions its behaviour on that representation (Godfrey-Smith 2013; Cao 2012, 2014; cf. Artiga 2016).

The first kind of case shown in Figure 4.4 is where a single representational vehicle R, with a range of states R_i correlating with a range of properties, acts as input to two distinct subsystems conditioning their behavioural outputs on the state of R (Case 1).

The two subsystems may be acting for different purposes and so may be making use of different correlations carried by R. The second kind of case is the converse: two different representations are made use of by a single consumer (Case 2). For example, R may correlate with colour and also with motion, where in some contexts behaviour should be conditioned on colour and in others on motion. R' indicates which context the system is in. In order to produce appropriate output, the organism conditions its behaviour on the conjunction of the state of R and the state of R'.

In the third and fourth cases, input influences output via more than one route. In Case 3 the two routes run in parallel. Like Case 2, the behaviour of the output subsystem depends on two different representational vehicles, but the state of the second vehicle is also dependent on the state of the first. In Case 4 we incorporate feedback: the state of vehicle R' is affected both by current input from R and also by input fed back from processing that took place on one of its own earlier states.

Real neural systems often include several of these elements at once. Below we pick out case studies that contain each in isolation. The aim is to show that none of them presents an obstacle to applying the UE approach.

4.5 One Vehicle for Two Purposes

A vehicle that carries correlational information about one state of affairs will usually carry information about many. Different downstream systems may be interested in different pieces of information: different correlations may be of use to each (Case 1 in Figure 4.4 above). There are many examples of that in animal signalling. A firefly produces flashes of light that signal its location to conspecifics. The signal means something like *receptive female here*. The signal also carries the information *there is a small nutritious insect here*. This second piece of information is capitalized on by predators.

In that simple case only one of the uses is part of a cooperative signalling system. Stegmann (2009, p. 873) gives an animal signalling example where both uses are at least partly cooperative. A chicken seeing a predator makes a distinctive call. This notifies conspecifics that there is a predator nearby so they can avoid it. It also notifies the predator that the chicken has seen it and could easily escape. Predator and prey share an interest in avoiding a pursuit if it will be unsuccessful. So, they share an interest in producing and acting on this signal. Conspecifics act on one piece of correlational information carried by the signal, potential predators on another.

When we turn to representations within a single organism, it is rarer that a representation is output to two entirely discrete consumer subsystems. Corollary discharge is perhaps one relatively clear case.[20] The signal sent to the motor system to drive action is also sent to perceptual mechanisms, which rely on it for the information it carries about what the animal is about to do. Very roughly, to the motor system it means *move thus and so* and to perceptual systems it means *I am actively moving thus and so*.

[20] Thanks to Rosa Cao for suggesting this example.

In Chapter 3 we saw that this second use of the motor signal, the efference copy, has an important role in enabling actions to be controlled smoothly.

Corollary discharge or efference copy is a very general principle of nervous systems, found even in the simplest organisms (Crapse and Sommer 2008). In more complex organisms like mammals it operates simultaneously at low levels (e.g. gating reflexes), and at higher levels (e.g. to allow predictive computations). A very simple example is found in the model organism *C. elegans* (widely studied because its nervous system has only 302 easily accessible neurons). When it senses pressure at the front, it produces a balancing motor response, which serves to stabilize its position. That reflex would get in the way when the animal pushes itself forward. So, the motor signal driving forward locomotion also serves to cancel the stabilizing reflex, freeing the animal to make forward progress. Thus, the corollary discharge signal both instructs the animal to move forward and informs the stabilization mechanism that the animal is going to move forward under endogenous control.[21]

That is a simple case of reuse, albeit one where arguably the two contents are of different kinds: one instructs that a world state be produced (directive content) and the other is designed to reflect the world (descriptive content). On another reading there are two kinds of directive content here: telling the motor system to push the animal forward and telling the stabilizing system to be inactive on this occasion. Chapter 7 deals with the question of what makes a content descriptive or directive. The importance of corollary discharge for now is that it shows how one vehicle can have two different contents deriving from two different downstream uses.

There are also likely to be other cases where the two contents are both descriptive, as with the chicken case, rather than one being directive and the other descriptive. Section 7.4 discusses a case where that arises in a slightly subtle way. The cases we will look at below are ones where it turns out that two different systems are using the same representational content, relying on it for different (but overlapping) purposes in different contexts.

4.6 Representations Processed Differently in Different Contexts

(a) *Analogue magnitude representations*

One potential case of the same vehicle meaning different things in different contexts comes from the analogue magnitude system.[22] It is used to represent numerosity, but it

[21] Where there is an efference 'copy' it may be that there are in fact two separate representational vehicles, one instructing the animal to move forward and a separate descriptive representation telling the stabilization mechanism that the animal is going to move forward. In cases where there is just one signal, then there will be a single representational vehicle which plays both these roles.

[22] I adopt the common label without making any claims about whether these representations qualify as analogue (rather than digital) in any useful sense; or about how best to draw the analogue–digital distinction and to characterize analogue computation.

seems to be capable of representing different things in different contexts: numbers of objects, tones, light flashes, and so on. I will conclude that this is actually a case where there is a common representation of numerosity. So the case will show how representations with a common content can be processed differently in different contexts.

Analogue magnitude representations correlate with the number of objects perceived in various situations: moving visual objects, static arrays of objects, tones, taps, flashing lights, and so on. There is very good evidence showing how the analogue magnitude system works in adults, infants, and non-human animals.[23] It can be used to compare numerosity across modalities; for example, judging if the number of tones heard is more or less than the number of objects in a visually presented array. However, the correlation between the analogue magnitude register and numerosity is imperfect. The further apart two magnitudes are, the more accurate subjects will be in comparing them (5 vs. 10 is easier than 5 vs. 6), but the more things there are to compare, the less accurate the comparative judgement (5 vs. 10 is easier than 15 vs. 20). That is, the representations follow Weber's law: discriminability is a function of the ratio of the difference between quantities to the overall quantity being compared.

There is evidence for one common register in the parietal cortex in which these numerosities are being recorded (Piazza et al. 2004, Nieder and Dehaene 2009). Registering numerosity in a common code affords ready comparisons across modalities and explains various priming and interference effects. Activation of this register R correlates with the number of items in the array or sequence presented, be they visual objects, flashes, tones, etc. Could R be a representation that has different contents for different downstream processes: numbers of objects in some contexts, numbers of tones in others, and so on? Or does R represent something common—numerosity—for all the uses to which it is put? Information about the types of item presented is not lost. Subjects can reach out to touch a flashing light, can follow moving objects with their eyes, can orient towards tones, and so on. So, there must be another component in the system somewhere with the functional role of telling downstream processing what kind of item it is dealing with, even if the number of items of that kind is recorded in a common register R. Simplifying considerably, let's suppose that contextual information about the kind of item is recorded in a separate register R'. Schematically, the set-up is the one we identified as Case 2 (Figure 4.5).

To home in on a task function, suppose people have been trained for monetary reward to report the number of items they have just been presented with. A visual array should be reported by pressing a button a corresponding number of times, and a sequence of tones is reported by moving a graduated slider on a screen. We can suppose that this input–output behaviour is a task function as a result of feedback-based learning. (The system will also have a more general task function as a result: to obtain money.) States R_1, R_2, etc. are states of increasing activation of register R, correlating

[23] Barth et al. (2003) in adults; Xu and Spelke (2000) in infants; Brannon and Terrace (1998) in monkeys. For reviews see Dehaene (1997) and Carey (2009, pp. 118–31).

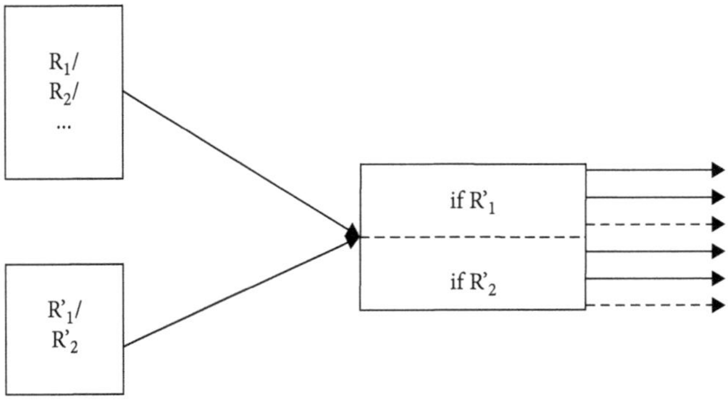

Figure 4.5 Case 2.

with the number of items just presented. State R'_1 correlates with the array being visual and R'_2 with it being auditory. The output behaviour is proportional to the activation of R, but the type of behaviour produced depends on whether register R' is in state R'_1, which leads to button presses, or in state R'_2, in which case the subject moves the slider.

When we look for UE information, this functional specialization is an important part of the algorithm that allows the system to perform its task functions. One register keeps track of item type and another deals with numerosity in general in a common register. That commonality is also crucial to the way the system is able to make cross-modal comparisons of numerosity. So, R comes out as carrying UE information about numerosity and R' about stimulus type. We could treat R as representing different contents for different uses: visual objects for some downstream uses, auditory events for others, and so on. However, recognizing that R is a common register for numerosity offers a more perspicuous explanation of how the system performs its tasks than if we were to treat it has having different representations for different downstream uses. So, the UE information approach suggests that R is simply representing the numerosity of the presented array.[24] It is a common representation which combines with another representation R' to generate different outputs in different contexts.

These kinds of considerations will often be at work in applying the UE framework to real systems. Where a register is deployed in a variety of contexts, that will push in the direction of its having a common content, one which abstracts away from particular

[24] There is a legitimate question here of what it is to represent numerosity, given that in many situations the domain being represented is discrete, whereas the vehicle of representation is either continuous-valued (e.g. a firing rate), or if it is discrete-valued (e.g. because it represents in terms of number of depolarizations, which are discrete events) then it has many more discrete values than there are integer values to be represented. Option 1 says that there are different values of the vehicle that all represent the same number of objects. Option 2 says that each state represents that the input has a certain non-integer magnitude (rational or real valued), and that it does so only approximately. How correctly or incorrectly it represents is given by the difference between the (real-valued) representational content and the (integer-valued) actual number of items; where degree of correctness can explain behavioural success and failure (e.g. the closer you are to getting it right, the more often the behaviour will be exactly appropriate to the number; and if appropriateness falls off in degrees, the more appropriate your behaviour will be).

sensory features of particular situations. That is, 'triangulating' on a common content, while not built into the framework, will often fall out of the explanatory considerations that underpin UE information. Perceptual representations will generally work like that. Being decoupled from any specific behavioural response also pushes in the direction of their having purely descriptive content, as we will see later (§7.4).

The analogue magnitude system also illustrates the idea that exploitable correlational information can be carried by a range of different states (§4.1a above). The activity of R varies and how active it is correlates with numerosity. That systematic relationship can extend to new cases. As a result of learning, R carries UE information according to a system that maps activity levels to numerosities: R_1 to one item, R_2 to two items, etc. Suppose that it happens that a numerosity of seven was never encountered during learning. R_7 nevertheless forms part of the same systematic relationship, so it carries the UE information that there are seven items present. Thus, when there is exploitable correlational information carried by a range of states in a systematic way, UE information can generalize beyond the instances that were encountered during stabilization (evolution, learning or contribution to persistence).[25]

(b) *PFC representations of choice influenced by colour and motion*

We are after a case where one and the same vehicle carries two kinds of correlational information, and intuitively one part of downstream processing makes use of it because of one kind of information that it carries, and another for another. In the previous example the functional specialization of the parietal cortex counted in favour of there being a dedicated common system for representing numerosity. So, we turn instead to the prefrontal cortex, which is less functionally specialized and carries information in a more domain-general way.

Mante et al. (2013) offer a model of information integration and context-dependent choice in the prefrontal cortex. Subjects (macaque monkeys) view an array of moving red and green dots (Figure 4.6). Sometimes the average motion is to the left, other times to the right, in each case with more or less coherence, making it more or less easy to judge the direction of coherent motion. Another dimension of variation is the proportion of dots of each colour: sometimes more red, sometimes more green. That discrimination is harder when the numbers of each colour are nearly matched. The task is either to judge the average direction of motion or to judge the preponderant colour. The task changes trial-by-trial, indicated by another stimulus (a yellow square or a blue cross at the bottom of the screen). The monkey responds by making an eye movement, either to a red circle on one side of the screen or to a green circle on the other side. When the 'colour task' stimulus is on, the monkey has to make an eye movement to the red circle if most dots are red, and to the green circle if most dots are green. When the 'motion task' stimulus is on, the monkey has to make an eye movement to the left if most dots are moving left, and to the right if most dots are moving right.

[25] As noted above (§4.1a), this mirrors a point made by Millikan (1984), which she describes as a kind of systematicity.

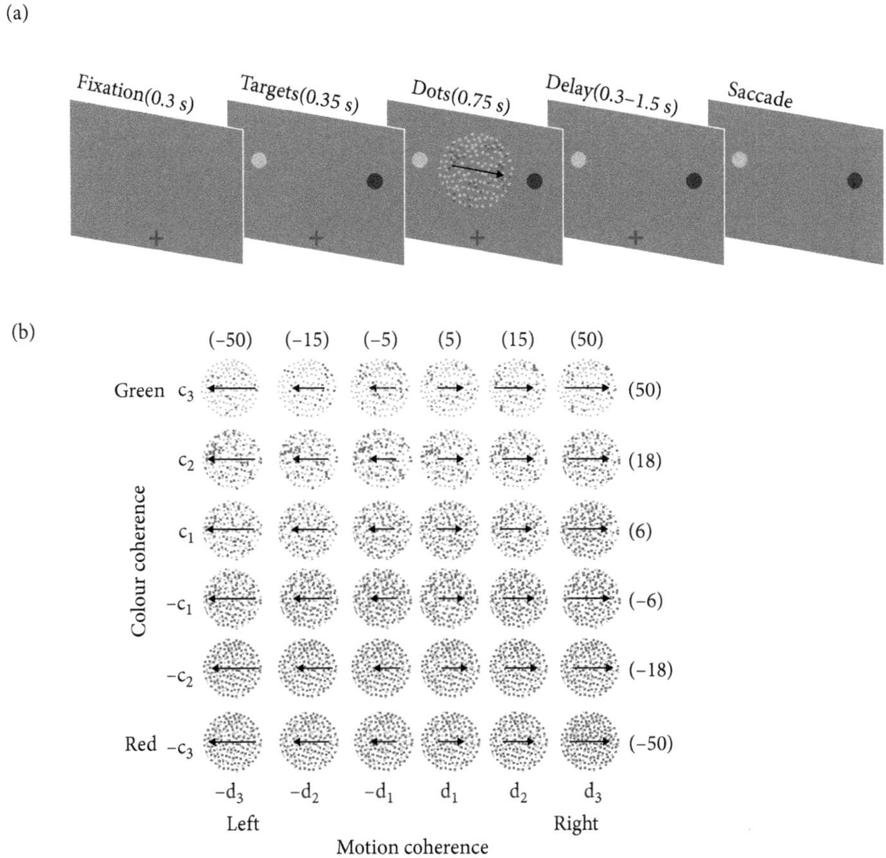

Figure 4.6 Behavioural task in Mante et al. (2013).

Mante et al. (2013) present neural and modelling evidence that the task is performed as follows. A network of neurons in the prefrontal cortex accumulates evidence about the majority colour and the preponderant direction of motion of the dots. We can think of this simply as having a representational vehicle that varies along two dimensions, one corresponding to colour and the other to motion. The graded nature of these dimensions allows for evidence accumulation. The longer the monkey looks at a stimulus, the more information it gathers about which is the preponderant colour and direction of motion. So, the activity along these dimensions increases with time, and does so more rapidly when the difference in the array of dots (of colour or motion, respectively) is more pronounced.

Activity in this network evolves over time towards one of two states, corresponding to making an eye movement in one of two directions. In the context of the motion task, evolution towards one action or the other is driven by evidence accumulated along the motion dimension. Information along the colour dimension is preserved (indicating the proportion of dots of each colour), but it has little or no effect on which choice is

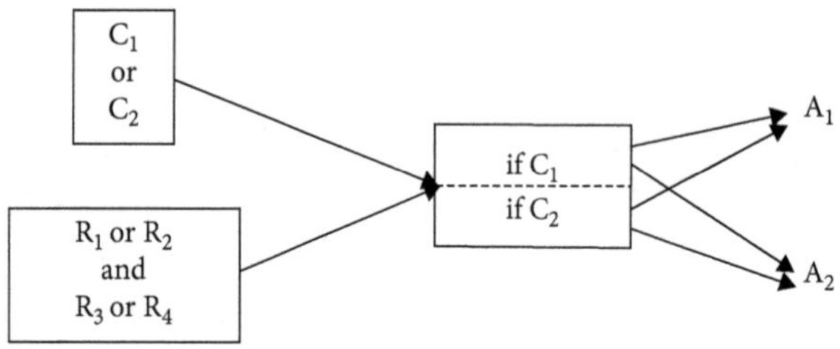

Figure 4.7 Schematic representation of the processing in Mante et al. (2013).

programmed. The reverse occurs in the context of the colour task: the colour-based dimension of variation drives evolution of the network towards a choice; motion-based information is preserved but non-efficacious. The contextual cue has the effect of selecting which dimension of variation of the representation will drive choice.

To capture the core of the case for our theorizing, let's ignore the graded activation and accumulation of evidence and consider the simplified processing diagram in Figure 4.7 (Case 2 again). We can treat the system as having two vehicles. C_1/C_2 correlates with the context cue indicating whether motion or colour will be the basis of reward in the current trial. The other vehicle can be in one of four states: R_1 or R_2 (colour) can each be paired with either R_3 or R_4 (motion). A processing step takes these states as input and produces A_1 or A_2 as output, corresponding to the two possible actions (simplifying slightly, suppose that these just program a saccade to the left or right, respectively). When in state C_1, R_1 vs. R_2 determines whether A_1 or A_2 is produced; R_3 vs. R_4 has no effect. The converse is true in state C_2.

As a result of learning, this system has the task function of obtaining juice. To find the UE information, we need to know what worldly conditions have to obtain for the monkey to get this reward. That has been set up in this case in distal terms, in terms of properties of the stimuli. The training regime was that the left/right decision is rewarded on the basis of the preponderant colour of the stimulus in one context and the preponderant motion in the other. Amongst all the correlational information carried by the components, the correlations which directly explain achieving these rewards are:

C_1: colour will be rewarded (if preponderant colour is red a saccade to the red circle will produce juice, if...)
C_2: motion will be rewarded (if preponderant motion is left a saccade to the left circle will produce juice, if...)
R_1: the preponderant colour is red
R_2: the preponderant colour is green
R_3: the preponderant motion is left
R_4: the preponderant motion is right

We could consider a different set of correlational information, the correlations with sensory inputs. C_1/C_2 correlates with certain sensory states (going with the yellow square or blue cross at the bottom of the screen); states of R also correlate with activity in the organism's primary sensory cortex. But an explanation of task functions in terms of this set of correlations would not be unmediated. It would have to be supplemented with further information about how neural properties relate to worldly properties. Then the correlations with the worldly properties would be doing all the explanatory work.

Lumping the states of C and R together into a single vehicle on which behaviour is conditioned would be less explanatory, for the same reasons we saw in the analogue magnitude case: it would overlook an important aspect of how internal processing manages to perform the task. Another alternative is that the system represents in a context-dependent way. In C1, it represents motion information but nothing about colour (those are mere correlations); in C2 the converse. But that overlooks the way those vehicle properties make a difference in the converse case.

Nevertheless, content attributions based on UE information retain some indeterminacy in this case. We already saw that there are two ways of capturing the UE information carried by components C: C_1 with *colour will be rewarded*, or with *if preponderant colour is red a saccade to the red circle will produce juice, if preponderant colour is green a saccade to the green circle will produce juice.* There is also indeterminacy at R, for example R_1 seems indeterminate between *the majority of the dots are red* and *the colour density in the middle of the screen is predominantly red*; or even *the moving surface in the middle of the screen is mostly red.* These collections of correlational information are equal candidates for explaining how the task of obtaining juice was performed robustly and stabilized by interactions with the environment. I would argue that finding this residual indeterminacy is the right result in this case.

4.7 One Representation Processed via Two Routes

The structure of Case 3 is illustrated again in Figure 4.8. Action is conditioned on two different representational vehicles, as in the previous section, but the second vehicle is also affected by the first. That is, the first representation affects behaviour via two routes.

Van Essen and Gallant (1994) produced an influential description of the primate visual system. One aspect of their account contains an example of the structure we are interested in (see Figure 4.9). There are several interconnections, and lots of

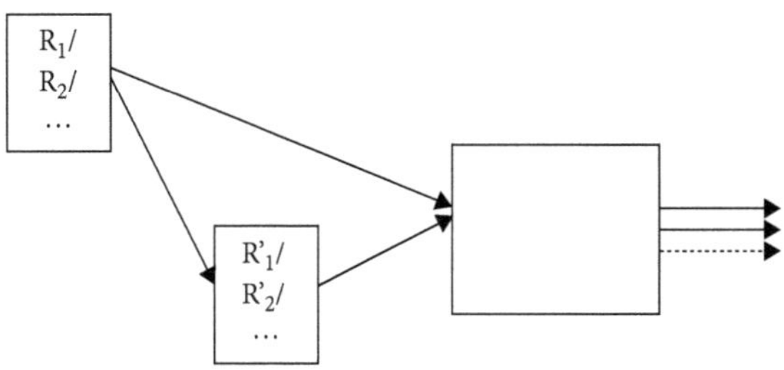

Figure 4.8 Case 3.

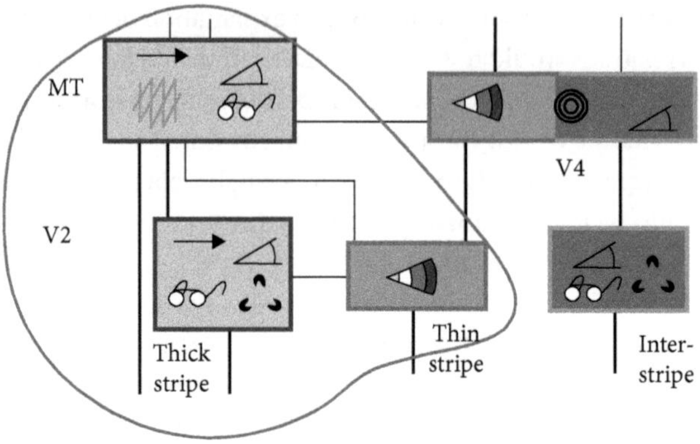

Figure 4.9 A detail from the visual information processing network described by van Essen and Gallant (1994).

connections into and out of the visual system that are not shown in Figure 4.9.[26] We restrict our attention just to processes occurring within the visual system, and amongst those just to the stages circled in Figure 4.9. The 'Thin stripe' area in V2 processes wavelength information, which is input directly to MT, and also affects processing in 'Thick stripe' V2 which in turn also affects processing in MT.

Van Essen and Gallant were primarily concerned to catalogue the functional connections in the visual system. Their claims about what information is being processed at each stage are more tentative and are a long way from being specified computationally. In order to have a concrete example to theorize about I will simplify and then fill in some details. That will give us a simplified but more specific case to work with.

Let us suppose that each box acts as a filter. Thin stripe V2 is tuned to colour dimensions, with different cells for different parts of visual space (we can think of this as retinal space). Area MT contains four different filters, but we can focus on just one, the one for plaid motion. These cells are sensitive to the direction of motion of surfaces in the visual field. They integrate local motion information and thereby correlate with the overall direction of motion of a presented surface. In some circumstances when observing gratings, that fusion can be broken so that the observer sees two superimposed gratings moving in different directions—so-called 'plaid' motion (Adelson and Movshon 1982; Burr 2014, pp. 766–9). We are supposing that this is in part because MT contains cells that are sensitive to the direction of motion of more than one surface in roughly the same portion of the visual field. In other conditions the observer will see just one moving grating.

This sensitivity to the direction of motion of surfaces is affected by chromatic information (Croner and Albright 1999; Thiele et al. 2001; Bell et al. 2014, p. 238). One route

[26] More recent work has altered this picture somewhat, but only serves to confirm that the feature we are interested in is indeed exemplified: not only is feedback vitally important, but it is also confirmed that visual processing does not occur in a strict hierarchy of subsystems or neural areas. Instead there are at least three streams that work in parallel, with interconnections between them (Shigihara and Zeki 2013).

shown by van Essen and Gallant (1994) is direct, from Thin stripe V2 direct to MT. When portions of nearby space have the same colour they are more likely to be treated as parts of the same surface. There is also an indirect effect of chromatic information, from Thin stripe V2 to Thick stripe V2 and then on to MT. Chromatic information affects the way Thick stripe V2 calculates the local direction motion. That in turn affects MT's calculation of where the surfaces are and how they are moving.

To theorize about content in this case I will simplify dramatically, so as to focus just on the dual route structure we are interested in. To give the mechanism a simple task function, suppose that the organism has been trained to reach out and intercept a moving object, doing so by tracking the direction of motion of observed surfaces. Then we can focus on one vehicle in MT, correlating with plaid motion direction, and suppose that it acts as input to the motor system so as to produce correlative reaching behaviour. That is a plausible task function if the organism has undergone training with feedback. Then the internal processing will have been tuned to enable the organism to catch objects stably and robustly.

Which set of correlational information, carried by internal vehicles, is directly explanatory of the system's ability to perform that task? The relevant MT activity correlates with the direction of motion of the surfaces of presented objects. (In our simplified setting it also correlates at output with reaching direction.) Thick stripe V2 has an array of vehicles each of which correlates with the local direction of motion of one portion of the visual field. The chromatic information in Thin stripe V2 correlates with many relevant properties; for example, with the wavelengths reflected by local areas, and with the colours of local areas. It is useful for this task because, when nearby areas have the same colour, they are likely to be part of the same surface. What is important, then, is the way activity in Thin stripe V2 for a local part of the visual field correlates with some property of presented surfaces that tends to be invariant for a given surface. Call that a chromatic-surface-property.[27]

In reality each of these components is involved in very many different task functions, and that will much more tightly constrain their contents. Even with our simplification, it is still clear that UE information will concern aspects of the distal objects the system is interacting with (e.g. motion properties), and features of the behaviour it performs on them (reaching direction). Most importantly, it is clear that the UE information will differ as between the three components we are considering. They are doing more than simply indicating *surface moving in such-and-such direction* with different levels of reliability at different stages. The problem of catching the object has been split up by having vehicles that track a series of relevant properties and performing a computation over those vehicles which is suited to calculating where to reach.

The consumer-based approach could bundle together the outputs of Thin stripe V2 and Thick stripe V2 and treat them as a single input, a vehicle which can be in a wide

[27] I won't divert into considering whether these properties are identical to colour properties. In any event, there may well be some indeterminacy here: a range of surface properties that Thin stripe V2 activity correlates with, each of which is an equally good candidate for explaining performance of this particular task.

range of states, and on which the output of MT is conditioned. The behaviour-relevant contents that can be ascribed to this conjunctive system are simply correctness conditions like *there is an object in such-and-such region moving in such-and-such direction*. Such contents offer no insight into how the system manages to compute the integrated motion of the surfaces of objects. It throws no light on the separate roles of wavelength information and achromatically driven local motion information in performing that computation. And it entirely overlooks the dual computations performed on wavelength information in solving the problem.

In short, this is another case in which the UE information approach does a good job of elucidating the way representational explanation works—and does so without having to appeal to a content-constituting representation consumer. Varitel semantics has no difficulty dealing with cases where a representational vehicle has a dual effect on behaviour via two different routes.

4.8 Feedback and Cycles

The final case involves feedback and cyclical information processing (Bogacz 2015). Rafal Bogacz describes a fully specified computational model of how the brain calculates the probabilities used to decide between a number of available actions (Figure 4.10). The model is far from being definitively established as the truth about how the brain generates this behaviour, but it is well supported by current evidence and it suits our purposes because it is specified in enough detail to get our theorizing off the ground.

The computation specified in Figure 4.10 calculates the probability that each available action is the best one to perform in the current content. When one of the probabilities reaches a threshold (determined by an attempt to maximize speed at a given accuracy), that action is performed. So, let us suppose that the $P(A_i)$ are inputs to a decision layer that detects when one of the inputs crosses a threshold and programs the corresponding action (added as a rectangular box in Figure 4.10). The computation shown calls for representations, not just of states of affairs, but of probability distributions over states of affairs. That is a new feature. Before getting into the details of the computation proposed by Bogacz, we need first to understand how the UE approach can apply to probabilistic representations.

A system can make use of the way its internal states carry probabilistic information.[28] In previous case studies we have only been concerned with the fact that a representation makes an individual world state more likely (R raises the probability that condition C obtains). But computations could make use of more fine-grained information carried by R: that when R is tokened, 75 per cent of the time the peanut is on the right and 25 per cent of the time the peanut is on the left, for example. When calculating what to do, the system can make use, not just of the fact that R raises the probability of some condition C1, but of the fact that when R is tokened, the probability that C1 obtains is p, the probability that C2 obtains is q, and so on. When this kind of fine-grained

[28] Shea (2014b) works out this idea in a canonical model of probabilistic population coding in the brain.

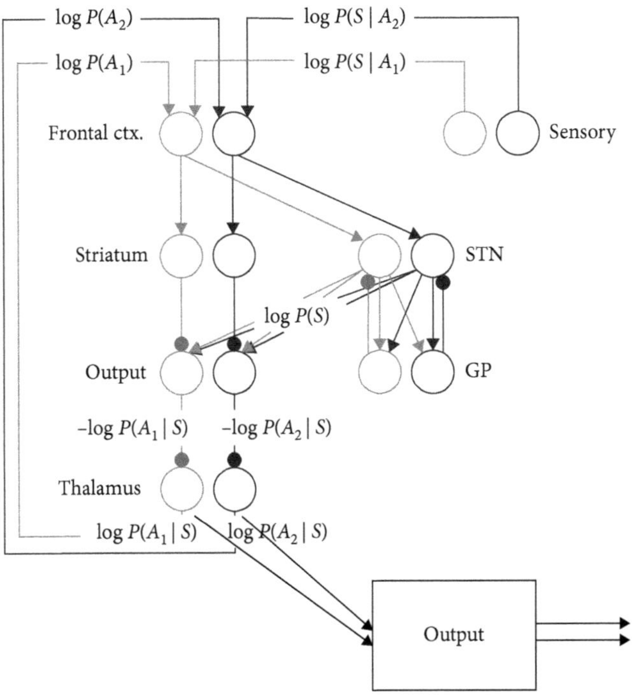

Figure 4.10 The neuronal computation proposed by Bogacz (2015) for deciding between available actions (adapted).

correlational information figures in explaining a system's performance of its task functions, internal states will end up having probabilistic representational contents.

To apply my account to these cases, all that's needed is a straightforward extension of the definition in §4.1a of exploitable correlational information carried by a range of states. A putative vehicle of content can be in one of a range of mutually incompatible states, so it counts as a random variable X. Now consider any item in the world that can be in a range of mutually incompatible states. It is another random variable Y. There is a joint probability distribution p(x,y) relating the two states. For every possible state of the representational vehicle, this gives the corresponding probability of each of the possible states Y. One way to think about p(x,y) is just in terms of frequency: fix on one particular state of the representational vehicle and ask how often Y is then in each of its possible states; repeat for each possible vehicle state.[29] A *fine-grained exploitable correlation carried by X about Y* is a joint probability distribution p(x,y) between X and Y that subsists for a univocal reason. (In defining X and Y as random variables it is implicit that each is constituted by states of items within delimited regions D and D', respectively. We are concerned with the probability distribution that subsists within those regions.)

[29] As before (§4.1a), this account depends on there being nomologically underpinned probabilities in the world, so these would have to be non-accidental, nomologically based frequencies; or they could be propensities or objective chances.

The joint probability distribution thus counts as a species of exploitable correlational information. The definition of UE information applies without any modification. A representational vehicle X enters into joint probability distributions with many different conditions in the world. When a system S encounters an object, the states of X might induce a probability distribution on possible sizes, possible directions of motion, possible categories of object, whether an object is animate or inanimate, and so on. For familiar reasons, X will also induce a probability distribution on more proximal facts like states of the retina and other brain states. All these probability distributions are candidates for UE information. Which qualifies depends on which figures in explaining S's performance of its task functions. For example, the probability distribution over motion directions may be directly relevant, given the way states of X are transformed in internal processing; probability distributions over states of the retina are less relevant.

When we were previously just concerned with probability raising, P(C|R) was important, and so was how much R changed the probability, i.e. how much P(C|R) differed from the unconditional probability P(C). That is also relevant here. Random variable X changes the probability of worldly states Y by comparison to the unconditional probabilities of states Y. This is measured by the mutual information between X and Y.[30] A set of conditions in the world about which a representational vehicle X carries more mutual information is, other things being equal, a better candidate to qualify as UE information carried by X.

We also need to generalize the idea of a correctness condition. The content represented by one of the values of X, x_1 say, is a probability distribution (call it $\hat{p}$) over world states Y: $\hat{p}(y|x_1)$. When x_1 is tokened its content would be completely accurate if $\hat{p}(y|x_1)$ exactly matched the actual probabilities of world states given x_1. When there is not an exact match we need a graded notion of accuracy/inaccuracy—of how close the content (represented distribution) is to the true distribution. There are various ways to measure how much the true distribution $p(y|x_1)$ differs from the represented distribution $\hat{p}(y|x_1)$. A standard measure is the Kullback-Leibler divergence. The Kullback-Leibler divergence of the true distribution $p(y|x_1)$ from the represented distribution $\hat{p}(y|x_1)$ measures how inefficient it is to assume that the distribution is $\hat{p}(y|x_1)$ when the true distribution is $p(y|x_1)$. It tells you how much more information (in bits) you would need in order to describe the true state of the world if you had represented it as $\hat{p}(y|x_1)$.[31]

[30] Taken on its own, the unconditional probability of Y is distributed across its possible states a certain way. That could be very indeterminate (e.g. all states of Y are equally likely) or already quite determinate (the unconditional probability of one or two states of Y is already quite high). This is measured by the entropy of Y, H(Y). Sharper distributions have lower entropy. States of X sharpen the distribution of Y, to a greater or lesser extent. That is to say, the entropy of the conditional distribution Y|X is less than the distribution of Y taken on its own (if X and Y are not wholly independent). This difference measures how informative X is about Y. So the mutual information between X and Y, I(X;Y) is given by the formula: I(X;Y) = H(Y) − H(Y|X).

[31] The Kullback-Leibler divergence is given by:

$$D\big(p(y|x_1)\,||\,\hat{p}(y|x_1)\big)=\Sigma_{y\in Y}\, p(y|x_1)\log\frac{p(y|x_1)}{\hat{p}(y|x_1)}$$

This is an appropriately graded notion of inaccuracy, going to zero when the world is exactly as it is represented to be.

Returning to Bogacz's model, the computational steps are as shown in Figure 4.10. The quantities are represented on logarithmic scales so that multiplication of quantities can be performed by adding firing rates. The system starts with prior probabilities for each of the A_i. It then gets sensory input S. It can then calculate how likely action A_1 is to be rewarded given S, $P(A_1|S)$, and so on. First it calculates $P(A_1)\,P(S|A_1)$, $P(A_2)P(S|A_2)$, etc. These then have to be normalized by dividing each by the sum of all of them so as to derive posterior probabilities for each action: $P(A_1|S)$ etc. So, the representations in frontal cortex are used in two ways. They go off to STN where they are summed, and the value of each is simultaneously preserved unmodified via the striatum, so that each separate value can be divided by this sum. If any of the $P(A_i)$ exceed the threshold at this point, the corresponding action is programmed. If not, these resulting $P(A_i)$ act as new priors for the next step of the calculation, performed on the next sensory input S.

For present purposes we are interested in the fact that information processing proceeds around a feedback loop before issuing in behaviour (Case 4, see Figure 4.11). The system has been tuned by learning to produce the action that is most likely to be rewarded given current sensory input, and to wait before acting in a way that gathers the optimal amount of sensory information to make a decision which optimizes a speed–accuracy trade off. It does so by tracking sensory information, processing it, and relying on that processing to program an action.

If the computational model above is well-supported, then it is likely to give the UE information carried by the components. According to the cognitive neuroscientists, organisms are acting near-optimally in a probabilistic environment in order to obtain the maximum amount of rewarding feedback. They look at how brain areas are probabilistically connected to states of the world in order to explain how the brain can be calculating appropriately—making calculations so that the behaviour will produce reward as often as possible. So, the test of the cognitive neuroscientific model's accuracy is the same as our test for UE information. If Bogacz (2015) is right about

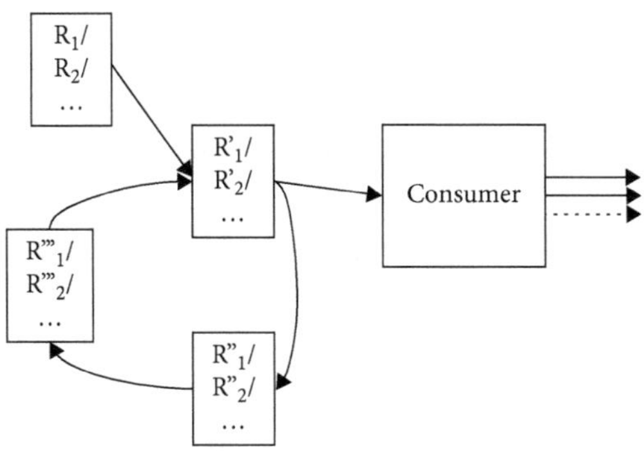

Figure 4.11 Case 4.

the correlational information carried by the brain areas he points to, and if he is right that neural firing is transformed in the way he suggests,[32] then his computational model is a good hypothesis about the UE information carried by these brain areas, and hence about their representational contents.

In short, the UE information approach does allow us to give a good account of why the components of this system have the contents they do, an account which in turn feeds into an understanding of why representational contents are suited to explaining behaviour. Here again, we had no need to appeal to a consumer system that plays a special content-constituting role. Internal processing involving feedback presents no obvious obstacle to applying the varitel framework.

4.9 Conclusion

Chapter 2 argued that representational content arises in many cases from the way relational properties of components of a system combine with facts about its internal processing. Taken together, internal processing over components standing in exploitable relations to features of the environment can amount to the implementation of an algorithm, an algorithm by which the system performs various input–output mappings. Turning this around, if we take a relevant input–output mapping, content is fixed by the exploitable relations carried by components which make the internal processing an implementation of an algorithm by which the system instantiates that mapping. Chapter 3 argued that task functions give us the input–output mappings that are relevant to content-determination. That was because of a cluster in which outcomes stabilized by natural selection, learning or contribution to persistence are also produced robustly and are generated by an algorithm that makes use of exploitable relations.

This chapter filled in the final part of the account, showing how correlational information counts as an exploitable relation within this framework. Correlations turn into content when they are exploited by a system—exploited in a very particular sense. Our definition of UE information pins down that sense: the content-constituting correlations are those which unmediatedly explain a system's performance of its task functions. We saw in this chapter how the UE approach fixes content in a range of case studies from cognitive science. It does so without having to appeal to representation consumers whose outputs play a content-constituting role. In each case study, contents fixed in this way do a good job of underpinning the characteristic explanatory grammar of representational explanation: correct representation explains successful behaviour and misrepresentation explains failure.

The next chapter argues that another exploitable relation also plays a content-constituting role, a relation in the ballpark of mirroring, isomorphism, or structural correspondence.

[32] I.e. that neural activity in these areas is added and subtracted in the way set out in the model—a way that is suitable to implement multiplication and division of quantities that are carried, not linearly, but on a scale that is related logarithmically to activation.

5

Structural Correspondence

5.1 Introduction

Organisms and other systems make use of exploitable relations between their internal states and the world in order to perform their task functions. The previous chapter looked at correlation as an exploitable relation. This chapter argues that structural correspondence is another exploitable relation. The existence of a structural correspondence, of an appropriate kind, is part of what makes it the case that certain systems represent as they do.

Cartographic maps act as a model for theorizing about structural correspondence. Spatial relations between points on a map correspond to spatial relations between locations on the ground. Plausibly, it is because the structure of the map mirrors the structure of the world that the map gets to represent the world. A relation in one domain corresponds to a relation in the other. More carefully, a structural correspondence is a relation-preserving[1] mapping from one set of entities to another. Points on the page of an atlas map to cities, and the mapping preserves spatial relations. When point

[1] More generally, structure-preserving. Here we focus on relation-preserving correspondence.

a in the atlas is closer to point b than point c, then that is also true of the cities to which they correspond.

A large body of work examines the idea that structural correspondence or isomorphism should be an ingredient in a theory of content. A central problem is to specify which kinds of relation can enter into the correspondence. Just as there is a very thin notion of property, a very thin notion of relation is also available. On the thin notion of property, any arbitrary set of objects corresponds to a property. For a relation, we can capture the entities that fall under the relation with a set of ordered pairs. The relation *taller than* is captured by the set of all the ordered pairs of people where the height of the first is greater than the height of the second. On the thin notion of relation, any set of n-tuples corresponds to a relation (an n-place relation). The problem for theories of content is that the thin notion of relation makes the idea of a structure- or relation-preserving correspondence extremely undemanding. If we instead require natural relations on either side of the correspondence, it becomes too demanding; and in any event something principled must be said about why some relations should be excluded and others should count. I will use the unqualified term 'relation' for the thin notion, and argue for restrictions to the candidate relations on both sides of the correspondence (representation, world).

It is a familiar point, but it is useful to recall why the existence of a structure-preserving mapping or functional isomorphism is a very liberal matter. This is illustrated in Figure 5.1. There are very many such mappings between any two sets of the same size. Suppose we want a representation of relation H on a set of entities x_i. H could be the hierarchical relation of dominance between a group of macaques. Take a set of putative representational vehicles v_i of the same cardinality. For any mapping I of the v_i onto the individual macaques x_i, there is a relation V on the v_i that corresponds to H: to see if V obtains between two vehicles, map them to the corresponding individuals under I, x_i and x_j, and see if x_i is above x_j in the hierarchy (i.e. see if H obtains between x_i and x_j). That will work even if we have fixed the mapping I from vehicles to macaques (v_i to x_i) in advance.

This liberality is one of the reasons why theorists have concluded that the bare existence of a structural correspondence cannot be the basis of content (Suarez 2003, Godfrey-Smith 1996, pp. 184–7, Goodman 1972; *pace* O'Brien and Opie 2004, Cummins 1989). From our perspective the problem is that most of these correspondences are not exploitable by the system in question. Our overall desideratum is to make sense of representational explanation. We do that by content being fixed by an exploitable relation between putative representations and the world, where the obtaining of that relation explains the system's performance of task functions. The bare existence of a structure-preserving mapping of the kind we have just seen is not something that will help a system perform a function. It is too insubstantial. So, our task is to identify a kind of structural correspondence which, when it obtains between vehicles and world, really amounts to an exploitable relation.[2]

[2] There are confusingly many relations in play. The exploitable relation is the correspondence, not the relations that are preserved under the correspondence.

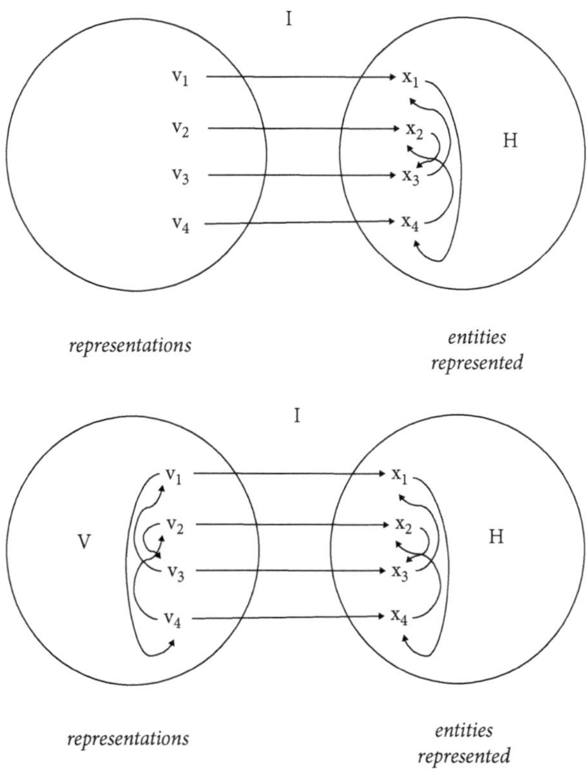

Figure 5.1 Whatever relation H we choose on the represented entities x_i (top panel), there is a corresponding relation V on the representational vehicles v_i (bottom panel).

We start with a case study: the cognitive map in the rat hippocampus (§5.2). This shows a substantive structural correspondence in action, enabling the rat to perform task functions. After some preliminary definitions (§5.3), I narrow down this more demanding substantive sense, *exploitable structural correspondence* (§5.4a). I go on to say what it is for one of these to be exploited (§5.4b). It then counts as an *unmediated explanatory structural correspondence* (UE structural correspondence). It will turn out that, in some cases where there is a superficially attractive structural correspondence, that correspondence is not being made use of (§5.5). These cases are contrasted with two further cases where an exploitable structural correspondence is exploited (§5.6). The final section (§5.7) discusses correlation vs. correspondence, approximate correspondence, and an evidential test for telling which structural correspondence is exploited.

5.2 The Cognitive Map in the Rat Hippocampus

One of the most important neuroscientific results in recent decades has been the discovery of 'place cells' in the rat hippocampus (O'Keefe and Nadel 1978, O'Keefe and Burgess 1996). Place cells are individual neurons whose firing is specific to the

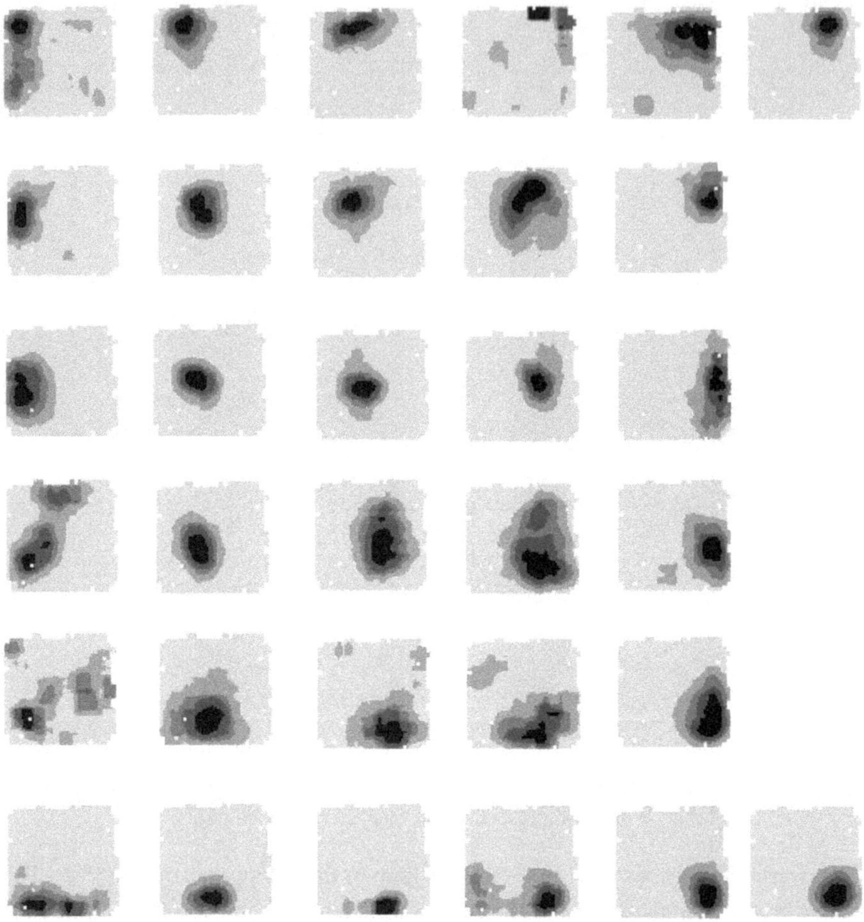

Figure 5.2 Firing rate plots for thirty-two place cells (O'Keefe and Burgess 2005). All the grey squares represent the same square arena, each showing where a particular place cell is active as the rat moves freely around the arena. Darker shading represents higher firing rates. Different cells are tuned to different locations.

animal's location in space. Figure 5.2 shows the firing pattern of an array of place cells. Each panel shows the sensitivity of a single cell, with shading showing the rat's location when that cell fires, darker shading representing more vigorous firing. So, there is one cell which is active when and only when the rat is in the top-right corner of the arena (describing it from an aerial perspective); another is specific for being half-way down the left-hand side, and so on.

The array of neurons as a whole gives a very accurate measure of where the rat is in the arena. That is a useful thing to have; for example, to learn by association that certain features (smells, objects, foods) are found in certain locations (Deadwyler et al. 1996). The correlational information carried by the place cell for the top-right corner could be relied on in instrumental conditioning to learn that, when at that location, the rat should pull a lever to obtain a reward. But notice that this is not to make use of a relation between the different place cells.

Indeed, taken on its own, the remarkable spatial sensitivity of the array of place cells does not depend on or give rise to any relation between the vehicles (cells). Nor are the cells spatially arranged in the hippocampus. They do not form a 'topographic map' like the retinotopic maps of visual space found in primary visual cortex. So, the remarkable discovery of the location-specific sensitivity of place cells does not, by itself, show that rats have a cognitive map. More recent work has shown however that there is an important relation on the place cells, the relation of co-activation. Cells corresponding to nearby locations tend to activate one another.

When the animal is at rest or asleep, firing of the place cells is taken offline; that is, it is no longer directly driven by input about the animal's current location (cp. stimulus-independence: Camp 2009). Offline activity shows characteristic sequences, corresponding to routes through space. 'Replay' occurs when offline sequences correspond to routes the animal has followed in the past (Wilson and McNaughton 1994, Foster and Wilson 2006, Diba and Buzsáki 2007). These connections could be built up associatively when sequences of place cells are active online as the animal explores an environment. 'Preplay' is also observed, where sequences of cells are active in advance of the animal moving, corresponding to a route the animal is about to follow (Dragoi and Tonegawa 2011, 2013). These preplay sequences lead to locations associated with reward, either because the animal has experienced rewards there in the past (Pfeiffer and Foster 2013), or because they can observe that food has been placed in that location (Ólafsdóttir et al. 2015).[3]

The current evidence suggests that rats use this prospective activity to plan the route they are about to follow. Let's suppose this allows them to select amongst possible routes, choosing a shorter one by selecting the shortest sequence of place cell firing. For simplicity, we can think of a process that activates several prospective sequences leading to a rewarded location and picks the shortest sequence as the one to follow. In fact, that search probably takes place in parallel across the whole array of place cells.[4] Either way, the co-activation structure over the place cells is being used as a proxy for

[3] There is parallel evidence in the human brain of similar kinds of preplay firing of sequences that correspond to trajectories through space (Horner et al. 2016, Bellmund et al. 2016); and for grid cells in entorhinal cortex, which also show prospective activity in the rat (de Almeida et al. 2012) which is coordinated with place cell activity at rest (Kropff et al. 2015).

[4] Most models envisage a diffusion process that starts at the place cell associated with reward and proceeds outwards in parallel across connected locations (Ponulak and Hopfield 2013, Khajeh-Alijani et al. 2015). E.g. Reid and Staddon (1998, 1997) have a model in which a value signal diffuses in parallel across an array of place cells, resulting in local signals of the direction of a shortest route to a goal (discussed by Godfrey-Smith 2013). Samsonovich and Ascoli (2005) construct a connectionist model in which relations of phase precession between place cells are used to search through routes in parallel, in a 'fan' proceeding outwards from the current location to all nearby locations. And Corneil and Gerstner (2015) construct an attractor network where associations between place cells constrain activity directly so that the offline preplay sequence spontaneously follows the shortest route to reward in the place cell activation space.

spatial relations between locations: this is a way of choosing an efficient route because place cells that co-activate each other correspond to locations that are close to each other in the arena.[5]

This case study fits squarely within the varitel framework. The animal has moved to a particular location *T* in the past and performed a behaviour there (e.g. pulling a lever and getting food). Its disposition to do that has been stabilized on the basis of feedback, because it received a food reward (let us say) at that location. It is then disposed and able to get to that location from a range of different starting points by a range of different routes (Pfeiffer and Foster 2013). Getting to *T* and getting food there have thereby become task functions for that individual. There is an internal-components explanation for performance of that task function: place cell activity makes use of correlational information about the current location and then proceeds offline in sequences that are driven by the co-activation structure. The animal picks an efficient route to a goal by picking the sequence that takes the shortest time to unfold during preplay. It then follows that sequence. That algorithm has been stabilized by learning in part because of a structural correspondence between co-activation on the place cells and spatial proximity on locations, relied on to calculate the route. That correspondence also feeds into an explanation of robustness, of how the animal manages to reach rewarded location *T* from a number of starting points by a range of different routes. In short, that structural correspondence is exploited. It explains the rat's performance of task functions. Therefore, I will argue, it is content-constituting: co-activation of place cells represents spatial proximity of locations.

In sum, the 'cognitive map' in the rat hippocampus illustrates how use of a structural correspondence to perform task functions can be the basis of representational content.

5.3 Preliminary Definitions

In this section I will say how I am using the terms 'structural correspondence' and 'structural representation', and what it is for a structural correspondence to play a role in constituting content. We start with structural correspondence. In all of our case studies the candidate to figure in the correspondence is some kind of relational structure. So, I define structural correspondence in terms of relations. It is a mapping under which relations are preserved.

On the world side, I will use x_i for entities and H for a relation between them. These are candidates to figure as representational contents. For example, a representation could represent that location-a is near to location-b. That would be to represent that a particular relation H obtains between two entities (locations) x_i and x_j. On the representation side, we need a way to talk about putative representations, since whether

[5] Notice that there is no straightforward consumer for the offline activity of a single place cell. Its activity has to interact with the activity of many other place cells. The result is then used, in conjunction with other inputs about current location, to condition behaviour.

they count as representations, and what they represent, flow from the obtaining of the correspondence relation. So I will call putative representations 'vehicles', v_i. V is a relation between the v_i. So the v_i potentially represent worldly entities x_i, and relation V's obtaining between v_i and v_j potentially represents that relation H obtains between x_i and x_j. For example, the activation of place-cell-a followed by place-cell-b potentially represents that location-a is near to location-b.

The toy example of a structural correspondence in Figure 5.1 was an isomorphism, a one-to-one mapping. There are the same number of worldly entities as there are vehicles, and every worldly entity is mapped to by just one vehicle. I follow others in relying on the slightly looser notion of homomorphism. A homomorphism allows two vehicles to map to the same worldly entity. There can then be representational redundancy: two vehicles can represent the same entity. So, there may be fewer worldly entities than there are vehicles. An isomorphism is a function from some v_i to some x_i and its inverse is also a function. A homomorphism is a function from some v_i to some x_i, but its inverse need not be a function. Finally, we are interested in a homomorphism that preserves relational structure.[6] Accordingly, I define structural correspondence as follows.

Structural Correspondence
There is a *structural correspondence* between relation V on vehicles v_m and relation H on entities x_n
iff
there is a function f which maps the v_m onto the x_n and
$\forall i,j\ V(v_i,v_j) \leftrightarrow H(f(v_i),f(v_j))$
(*mutatis mutandis* for other polyadicities[7])

There is an issue here about structural representations and their parts. A map is a structural representation and its parts are also representations. One part of a map might consist of two points separated by 6.5 cm, representing that Cardiff is 65 km to the east of Swansea. A point taken alone can also be a representation (e.g. of Cardiff—an unsaturated representation). The definition of structural correspondence above does not require this. The vehicles taken alone need not be potential representations. This would allow for a representational icon whose parts are not themselves representations. The icon would represent in virtue of a structure over vehicles, and those vehicles would be parts of the icon, without also supposing that the individual vehicles will qualify as representations in their own right. I don't want to take a stance on

[6] If the homomorphism is not an isomorphism, then the relation H between worldly entities needs to be reflexive, at least for entities that are mapped to by two different vehicles. If the structural correspondence maps v_i and v_j to the same x_k, then relation H has to obtain between x_k and itself. For example, relation H could be *being less than* 5 *cm away*.

[7] The definition can readily be generalized to cover any collection of relations and operations, of any polyadicities, following the mathematical definition of a relational homomorphism (although the latter are usually thought to range over mathematical objects).

whether this is possible. In all of our case studies the parts are also representations. So, I will define terms that way, following the standard definition, to stop the language becoming unbearably complex. For now, I just note that my approach could in principle apply to structures whose parts are not themselves representations.

The standard definition of structural representation does take the parts to be representations. What it takes to be a structural representation is that a relation on the representations represents a relation on the entities represented (Ramsey 2007, pp. 77–92; Swoyer 1991; Shagrir 2012). For example, spatial relations between points on a cartographic map don't just correspond to, but represent, spatial relations between the locations picked out by those points. That is a first-order resemblance, but any relation on a set of representations could in principle represent a corresponding relation on the entities represented. The obtaining of a relation between two vehicles represents the obtaining of a relation in the world. The obtaining of the relation of *being within 5 cm of* between two points on a map represents, for example, that the relation of *being within 5 km of* obtains between two cities in the world.

> *Structural Representation*
> A collection of representations in which a relation on representational vehicles represents a relation on the entities they represent.

We are in fact interested in something slightly stronger than there being a structural representation. Our question is whether something's being a structural representation is based on the obtaining of a structural correspondence: whether structural correspondence is partly constitutive of content. One could set up a convention in which a structure happens to represent structure in the world, but then it would be the convention rather than the existence of the correspondence which is constitutive of content. For example, I could make a list of names and stipulate that the relative size of the fonts represents the relative heights of the people named. Then there is a relation on the vehicles (relative font size) which represents a relation on the people (relative height)—so it fits the definition of structural representation—but the structural correspondence is not what is fixing content. The case studies in this chapter are stronger. The structural correspondence does fix content: the existence of a certain kind of structural correspondence is part of what makes it the case that a collection of vehicles are representations with a certain content.

> *Structural Correspondence as Content-Constituting*
> A structural correspondence I is *content-constituting*
> iff
> the existence of structural correspondence I, of an appropriate kind, between a relation V on putative representations v_m and a relation H on the entities x_n represented by the v_m is partly constitutive of V on v_m being a structural representation of H on x_n.

The varitel framework applies to cases where content arises out of a system's making use of exploitable relations. So, if a structural correspondence is going to be content-constituting, it has to be used by the system. In order for it to use a structural correspondence, the system has to be sensitive in some way to the relation V between vehicles. That relation has to make a difference to downstream processing, and ultimately to the behaviour produced.

For contrast, consider the vervet monkey's system for signalling the presence of predators. Vervets make three types of alarm call for three types of danger: say R1 for aerial predators like eagles, R2 for ground predators like leopards, and R3 for snakes (Seyfarth et al. 1980). Conspecifics hearing the call make use of the fact that R1 correlates with eagles in order to behave appropriately; similarly for R2 and R3. This is a classic case of making use of correlational information. But notice that there is also a one-to-one mapping between representations and their correctness conditions (R1 to *there is an eagle*, and so on). And as ever, lots of relations are preserved by that mapping. Let's focus on just one: how high off the ground the predator is usually found. Eagles are higher up than leopards, which are higher up than snakes. I pick that arbitrarily, just to make a point. There is no evidence that the calls are telling macaques anything about height off the ground.

So, we are concerned with a relation H, *higher than*, between worldly entities. H applies to just the following ordered pairs: <eagle, leopard>, <leopard, snake>, <eagle, snake>. Now, as we've seen, there will be *a* relation on the vehicles (alarm calls) that corresponds to H. Call it V. V applies to just the following ordered pairs: <R1, R2>, <R2, R3>, <R1, R3>. So, there is a structural correspondence between relation V on the alarm calls and relation H on the predators. However, the existence of this structural correspondence is of no significance to the vervets. They are not making use of it as they process the alarm calls. They are not sensitive to whether relation V obtains between the calls. Vervets have evolved to respond appropriately to calls that have the acoustic features of R1, but that does not depend on comparing R1 to R2 or R3, or making use of any relation between R1 and the other calls. The system exploits the correlations (between R1 and eagles, R2 and leopards, and R3 and snakes), but does not exploit the structural correspondence between V and H. The structural correspondence is not content-constituting. Nor is it a case of structural representation: the relation V on the alarm calls, defined above, does not represent *higher than* (or anything at all).

The requirement that a structural correspondence be used in order to be content-constituting allows us to cut down very considerably on the problematic liberality of structural correspondence. To be content-constituting, a structural correspondence has to be exploitable. In the next section I pick out a class of structural correspondences that are candidates to be exploited. This is the restricted notion we needed: it avoids wild liberality and is restricted in a principled way. I go on in §5.4b to say what it takes for a substantive structural correspondence of this kind to be exploited, hence to constitute content.

5.4 Content-Constituting Structural Correspondence

(a) Exploitable structural correspondence

Recall that theories of content are faced with the problematic liberality of the general notion of structural correspondence, and hence need a more restricted notion that is substantive and well-motivated. If a structural correspondence is going to figure in varitel semantics, it has to be usable by the system, something that is a candidate to explain the system's performance of task functions. This section spells out that substantive sense. I label it 'exploitable structural correspondence'. (This is not going to be circular: it is not defined in terms of being exploitable.)

In the rat navigation case, the relation of co-activation on place cells (representations) was something that processing was sensitive to and was used in processing. That relation of course corresponds to very many relations in world, but it is the correspondence with the relation of spatial proximity on locations that makes sense of how the animal manages to perform its task functions. Spatial proximity between places is directly relevant to the task of following shortest routes to reward.

When we examine this privileged, content-constituting structural correspondence, on one end it has a relation that downstream processing is sensitive to, and on the other it has a relation in the world that is of significance to the system, given the task functions it is called on to perform. Although there being a structural correspondence is only a very weak requirement, that there should be a structural correspondence of this kind is very demanding indeed. It is a considerable achievement to have place cell activity organized in this systematic way—having a correspondence that the animal can make use of. This case exemplifies what it is for a structural correspondence to be exploitable.

> *Exploitable Structural Correspondence*
> An *exploitable structural correspondence* is a structural correspondence between relation V on vehicles v_m in a system S and relation H on entities x_n
> in which
> (i) V is a relation that processing in S is systematically sensitive to; and
> (ii) H and x_n are of significance to S.

Significance to S is significance relative to the way outcomes produced by S are stabilized and robustly produced. Sensitivity is also system-relative. Processing in rat hippocampus is sensitive to patterns of co-activation between place cells. It is not sensitive to the colour of the cell bodies of the place cells. Nor is it sensitive to where within a layer of the hippocampus the place cell happens to be located: connectivity not location of the cell is what counts. Primary visual cortex is structured retinotopically: the spatial arrangement of neurons corresponds to the spatial lay-out of retinal areas they respond to. However, the significance of the retinotopic organization is debated (Chklovskii and Koulakov 2004, Knudsen et al. 1987). A central issue in that debate is precisely whether downstream processing is systematically sensitive to the spatial arrangement of neurons.

For place cells, the exploitable structural correspondence only exists because associationist learning has built up a co-activation structure. Only after that does the relation between vehicles—one place cell firing immediately after another—correspond to a relation between places which qualifies it as an exploitable structural correspondence (because only then does it have a relation on the world end, spatial proximity, which is of significance to the system). One might argue that the location-specific sensitivity of place cells is very useful even before this learning has taken place. After all, that is what allows simple associationist learning to construct a cognitive map. I don't want to resist the idea that there is something exploitable in a broad sense even before the co-activation structure exists—the rat already has something useful. However, I use 'exploitable structural correspondence' in a specific sense: it requires that a relation between vehicles already exists which downstream processing is sensitive to.

There is a danger of getting confused here amongst the various relations in play. The exploitable relation is the structural correspondence. The relation of co-activation between place cells is a different relation, a relation on one side of the structural correspondence. It is not itself the exploitable relation.

Downstream processing has to be sensitive to a relation between vehicles if that relation is to form part of a structural correspondence which is exploitable by the system. Neural processing is certainly sensitive to relations between firing rates. In many cases it is also sensitive to fine-grained differences in the exact time that spikes are produced in different neurons. There are debates about whether some neural computations use a phase code, that is a code in which what counts is the time when a neuron fires in relation to a background oscillatory rhythm in the population of neurons. If so, phase differences are also candidates for the relation (V) on the representational side of an exploitable structural correspondence.

Plasticity can drive changes in the sensitivity of downstream processing. Then a relation between vehicles which previously did not count, because downstream processing was not systematically sensitive to it, may turn into a candidate. In some cases it is feedback from stabilization that drives this plasticity. In that case the exploitable structural correspondence becomes established at the same time as it contributes to stabilization. So the exploitable correspondence that is made use of to perform a task function need not pre-exist the stabilization process which grounds the task function. As just mentioned, a wider notion of exploitability is available, which does not require that the system is yet sensitive to the relation between vehicles. The category of *potentially exploitable structural correspondence* covers cases where a system can readily adjust so as to make downstream processing sensitive to a relation between vehicles, or can readily put vehicles into a relation (like co-activation) to which downstream processing is already systematically sensitive. It may well be important that some systems have access to many potentially exploitable structural correspondences. The definition of exploitable structural correspondence is narrower, however, because the aim is to home in on a content-constituting correspondence. We are concerned with the actual sensitivity of the system, as configured. The class of potentially exploitable relations may in any event be less well-defined.[8] Potential exploitability certainly comes in degrees.

The definition of exploitable structural correspondence also requires that relation V should make a *systematic* difference to downstream processing. What this amounts to will depend on the types of processing in question. The general idea is that V should have downstream effects that operate according to common principles. So, when the same relation obtains between different pairs of vehicles (co-activation of two place cells), downstream processing should do the same thing in each case (treat this as a single step in calculating routes). If V comes in degrees, then processing should be systematically sensitive to those degrees. For example, if V is a difference in the time of firing, then there should be a systematic relation between the way downstream processing treats differences of 1 ms, 2 ms, and 3 ms. One way of spelling this out is to say that V should figure as a projectable property in a special science law describing the processing of the system. Whether that is the right way to understand systematic sensitivity is an issue about causation for philosophy of science more generally and not a proprietary problem for theories of content. In order not to pre-judge that issue, my definition simply makes use of the idea of systematic sensitivity, which is a resource needed throughout the sciences.

Turning to the other end of the correspondence, the things in the world being represented, the definition requires that the correspondence should be with entities x_n and a relation H that is of significance to the system. What counts as significant for the system is relative to its task functions. In the cases we are considering significance to the system narrows the candidates down to natural objects, properties, and kinds in the world. But I don't need a general account of what naturalness amounts to: the significance requirement imports a system-relative constraint (which will require naturalness in many cases). As a result, it will mostly cut out gruesome and disjunctive properties as candidates for content, but does so in a system- or organism-relative way.

Notice that there are different constraints on the two sides of the correspondence. An obvious restriction would be to introduce a naturalness constraint on both sides of the correspondence. But any restriction needs to be well-motivated. The motivation provided by the varitel framework calls for system-relative restrictions on both sides of the correspondence but different ones. On the vehicle side, the restriction is motivated by the role of inter-vehicle relations in downstream processing. On the world side, the restriction is motivated by whether relations in the world are significant for the

[8] There is a parallel here with exploitable correlations. It is useful to have a system that can create exploitable correlations by building associations between existing correlation-carriers, e.g. a new C that is active only if A and B are. The existing correlation-carriers A and B give the system the potential to track C. Only once the new correlate is created, however, is there a new exploitable correlation. (And it is still a further step, of course, for the system to make use of that exploitable correlation.)

system (significant for its performance of task functions). That is why our exploitable structural correspondence has different restrictions on each side.

On occasions when an exploitable structural correspondence is being used, relation V is instantiated between some actual vehicles, and relation H is instantiated between some actual things in the world. When I talk of a structural correspondence being instantiated, what I mean is that an instance of the relation V is instantiated between two vehicles, together with an instance of relation H being instantiated between the two worldly entities to which they correspond.

So far, we have seen the following: out of all the very many structural correspondences that exist, there are some that are ready to play a role in explaining task functions. In these cases it is a substantial achievement to have such a correspondence in place. This is the sense in which the Survey of India was such a major achievement (and such a powerful tool of colonial control). Why bother? After all, the haphazard distribution of pebbles on Horse Guards Parade already bore *a* structural correspondence to the settlements, mountains, and rivers of India (under certain mappings). What the Survey achieved was to create an artefact bearing a relation that users are readily sensitive to (spatial separation on a sheet of paper) which corresponds to a relation on the ground of significance to the colonial regime (distance). An exploitable structural correspondence is useful to have and an achievement to create. Next, we see what it is for an exploitable structural correspondence to be made use of in performing task functions: to be an 'unmediated explanatory structural correspondence'.

(b) Unmediated explanatory structural correspondence

Our desideratum was to make sense of representational explanation, and the varitel framework achieves that by making content a matter of exploitable relations which explain performance of task functions (the explanandum spelt out in §4.2a). So, an exploitable structural correspondence constitutes content when it explains how certain outputs F_j were stabilized by one of the feedback processes in Chapter 3, and/or how they were robustly produced.

> *Unmediated Explanatory Structural Correspondence*
> A structural correspondence I between relation V on vehicles v_m in a system S performing task functions F_j, and relation H on entities x_n, is a *UE structural correspondence* iff
> (i) I is an exploitable structural correspondence; and
> (ii) instantiation of I plays an unmediated role in explaining, through v_m and V implementing an algorithm, S's performance of task functions F_j[9]

[9] As before (§4.2a), being 'unmediated' is intended to rule out cases where I is explanatory because its targets fall under another structural correspondence I* with further objects and properties that are the ones which figure in a causal explanation of stabilization and robustness.

I argued at the end of the last section that the rat's algorithm for picking shortest routes exploits the structural correspondence between co-activation on place cells and relations of proximity between the places to which they are sensitive. Consider a location *T* at which the rat has previously experienced a food reward. It can get back there from a variety of starting positions by a variety of different routes, so getting to *T* is a robust outcome function. Getting to *T* is a stabilized function of the system because getting to *T* in the past (and doing something there) led to getting food, a type of feedback which reinforced the disposition to go to *T*. This outcome probably has an evolutionary function as well, deriving from the evolutionary function of the whole spatial navigation and learning mechanism. Getting to *T* and getting food there are also stabilized functions in virtue of contributing to survival of the organism. So, getting to *T* clearly meets the conditions for being a task function.

The structural correspondence between place cell co-activation and spatial relations figures in explaining how the rat gets to *T* robustly and how doing so was stabilized. Another part of the story is the UE information carried by the place cells when the rat is moving and they are online. That allows the rat to navigate from different starting points and to register when it has reached its target. Another important piece of the story is that the system carries correlational information about the location of previously encountered rewards. These correlations come together with the structural correspondence to explain how reaching *T* was stabilized by reinforcement learning. So, this is a case of UE structural correspondence. The need for convergence between UE information and UE structural correspondence is an important source of the determinacy of both kinds of content in these cases.

The final step is much shorter. UE structural correspondence is a sufficient condition for having content:

> *Condition for Content based on Structural Correspondence*
> If there is a UE structural correspondence between relation V on vehicles v_m and relation H on entities x_n
> then $V(v_i, v_j)$ represents condition $H(x_i,x_j)$

The existence of an exploitable structural correspondence is a necessary part of this sufficient condition for content. So, according to this theory, structural correspondence (of an appropriate kind) is content-constituting.

The sufficient condition for content is formulated so as to be neutral between descriptive content (*H obtains*) and directive content (*bring about H*). That distinction is discussed further in Chapter 7. The structural correspondences discussed in this chapter all underpin descriptive contents: so when the relation V is instantiated between two vehicles v_i and v_j, this represents that the relation H *obtains* between two corresponding entities. For example, when two place cells are co-activated, this represents that the corresponding locations are close to one another in space. (Which location each cell is representing is fixed by UE information.)

My overall approach may generate a suspicion of circularity. We want exploitable structural correspondence to be a resource that can be made use of, but the terminology suggests that being usable is what makes something an exploitable structural correspondence in the first place. In fact, exploitable structural correspondence is not defined in terms of being exploitable, but in terms systematic sensitivity and significance for the system. Nor does the definition of UE structural correspondence mention exploiting a relation. So there is no definitional circle.

In most of our examples, content of the vehicles v_m has been fixed independently of the relation V over them. Names on a map represent towns and cities by convention. Rat place cells represent locations because they correlate with locations and are used to perform appropriate actions at those locations. However, there can be cases where which entities are represented and which relations are represented are both fixed in parallel. Think of a cartographic map with unlabelled points at locations (see Figure 5.3). Arguably, each point refers to a particular location. The fact that a point on the map refers to a specific location is fixed by the spatial relations of that point to other things on the map (e.g. to other points, a coordinate grid, and/or a benchmark).[10] Similarly, we could imagine introducing a new place cell into the co-activation structure of the hippocampus, but without its having any online sensitivity to location. That cell would acquire a content—would represent a location—in virtue of its co-activation relations to other place cells. So, a UE structural correspondence can determine content about entities (x_n) and their relations (H) all at once.

Figure 5.3 A simple map. Notice that the unlabelled points pick out locations. They do so in virtue of their spatial relations to other things on the map.

[10] I won't attempt a careful treatment here of the appropriate compositional semantics for maps, e.g. whether absence of a symbol at a location represents absence of the corresponding property instance at that location. See Blumson (2012), Camp (2007), Rescorla (2009b, 2009a).

Another way a new exploitable structural correspondence can come into existence is by learning new relations on existing entities. We saw an example of that with the co-activation structure on place cells. A very different example of that is learning the sequence of count words. Taken as phonetic patterns, the count words 'one', 'two', 'three', and so on are merely arbitrarily related. There is *a* relation on them which corresponds to the mathematical relation of successor, but for the child who has not learnt to count, that relation has no significance. Learning the count sequence by rote, however, gives rise to a new relation on these phonetic patterns. Once memorized, activating one in auditory-motor imagery tends to activate the next one in the count sequence. The child then has a relation it can make use of in downstream processing.[11] This is another way that a new exploitable structural correspondence can be established over a set of putative representations, in this instance not by changing the sensitivity of downstream processing, but by altering the structures that exist over the representations.

At the personal level, mnemonics are a common way that we create structures that we can then use in reasoning. Once I've learnt a mnemonic for the first eight US presidents (will a jolly man make a jolly visitor?), I can use it to calculate temporal relations: van Buren came after Jackson and a long time after Washington. When a memorized sequence becomes automatized, like the count-word sequence, it may become possible for automatic and non-conscious processing to make use of the correspondence. And there are doubtless many cases like the rat place cells where sub-personal learning processes produce a co-activation structure that can then be used for the way it corresponds to objects and properties in the world (as in §5.6b below).

So, an organism will typically have the potential to create very many different exploitable structural correspondences, more or less easily in different cases, by constructing new relations on representational vehicles or by making downstream processing newly sensitive to an existing structure on vehicles. Such changes take a potentially exploitable structural correspondence—a category that comes in degrees and we only need to gesture at loosely—and turn it into an exploitable structural correspondence, which we have defined precisely above (§5.4a). When an exploitable structural correspondence is used to perform task functions, either when it is constructed or subsequently, it becomes a UE structural correspondence, thereby constituting content.

5.5 Unexploited Structural Correspondence

This section goes a bit deeper into the question of which structural correspondences count and which don't. In the rat navigation case an exploitable structural correspondence is exploited—it plays an unmediated role in explaining the rat's performance of

[11] This learnt relation plays an important role in Carey's theory of the acquisition of number concepts (Carey 2009; see also Shea 2011c).

task functions. I start by contrasting that with a case where an obvious structural correspondence is not in fact being exploited and is not part of the basis of content.

With familiar public representations, when there is an obvious structural correspondence, we often use it. Indeed, the correspondence is often set up because of its ease of use. That is why maps use space as a representational vehicle. Many other ways of displaying data in charts and graphs also rely on space as a representational vehicle, using spatial relations to represent a very wide range of relations in the world (e.g. genetic relatedness, age, income, . . .).

Another common relational coding is colour. Colours are used on weather maps to represent temperature and on fMRI brain scans to represent blood flow. Relations between different regions are visible at a glance. Colours are also used in non-spatial ways of arranging data. A list of students in a class might be colour-coded by their recent test score along an axis from blue for low scores through green, yellow, and orange to red for the highest scores. Taken alone, that coding just associates a piece of correlational information with each name in the list. But the colours make it easy to compute using relations between the test scores; for example, to sort the class into three groups with similar scores, or to sort students into pairs with very different scores. These ways of using the data make use of the exploitable structural correspondence between colour space (on the representations) and relative test scores (of the individuals represented).

When the users are people, there is not much of a gap between a structural correspondence being obvious and people starting to use it. In cases in cognitive science, however, it is relatively common that an obvious structural correspondence, even one that the system could easily become sensitive to, is not exploited. As I have argued previously, the structural correspondence that exists in the honeybee nectar dance is not being exploited in my sense (Shea 2014a, pp.128–30). Although there is an obvious relation between different dances, consumer bees are not making use of relations between different dances in deciding where to fly. As standardly described, the behaviour does not take more than one dance as input. Nor does the relation between dances enter into computations in other ways. This is a case of UE information but not UE structural correspondence; indeed, it is not a case of structural representation. There is no content-constituting structural correspondence. Condition (i) on being an exploitable structural correspondence in my sense is not met (§5.4a)—downstream computations are not sensitive to the putatively representational relation on the representational vehicles.

The bee dance has a different property which is important, and worth dwelling on briefly. There are different dances for different directions, and it is not arbitrary which dance goes with which direction. There is a systematic relation between dances which mirrors the systematic relation between directions. The system of available signs exhibits what Godfrey-Smith has called 'organization' (Godfrey-Smith 2017, p. 279). Contrast a nominal sign system like the count words for numbers. Whether a sign system counts as organized or nominal depends on what qualifies as a systematic relation between signs, and on which relations in the world are candidates to be mirrored. Does the systematic relationship between signs need to be a natural relation? Does the relation mirrored also need to be a natural relation? This is similar to the question of what counts as an exploitable structural correspondence (§5.4a), but I won't attempt to resolve it here for the organized-nominal distinction.

We saw in the last chapter that often correlational information is not carried pointwise, representation-by-representation, but is carried systematically by a range of representations about a range of states (see the definition of 'exploitable correlational information carried by a range of states' in §4.1a). That is important because it allows a compact mechanism to deal with a large number of different syntactic items representing many different directions of nectar (potentially continuum-many). It extends to new cases, beyond those on which it was stabilized, when they fall into the same system. It also makes the system error-tolerant, since a representation that is incorrect but approximately true will prompt behaviour that is close to being appropriate to the situation (flying off in roughly the right direction). When UE information is based on correlational information about a range of states, the need for a systematic account that applies to a range of different representations will effectively cut down content indeterminacy. So, organization, when it exists, is an important part of the way a system of representations does its job.

Organization is sometimes assimilated to structural representation, but they are distinct phenomena. Organized signs are tokened on different occasions, during different behavioural episodes. The relation between signs is useful because the different occasions are related in a systematic way (e.g. the behaviour called for is systematically related to the direction in which nectar is located). Structural representations have parts that are tokened together during a single episode of behaviour. The structure allows the organism to behave in a way that is appropriate to the occasion. A structural representation is a single representation with representational parts; an organized sign system is a series of different representations. A structural representation must have semantically significant constituent structure; a sign in an organized sign system need not.

The parts of a structural representation need not be tokened at the same time in order to count as parts of the same representation. Parts tokened at different times can be used to calculate what to do on a single occasion. For example, place cells are activated one after the other. Their activity need not overlap in time. This is also a feature of Robert Cummins's well-known example of the driverless car (Cummins 1996, pp. 94–5; see also Ramsey 2007, pp. 198–9). The car's wheels are steered by a pin driven along a slit in a card (see Figure 5.4). When the slit is to the right of centre, the wheels are steered to the right and the car turns right (the converse for left turns). If the car is placed in a track whose turns match the card in the right way, it will follow the track without hitting the sides. Although it looks like there is a standing structural representation of the environment (the card), the way representations are tokened so as to drive behaviour is by the pin being located at different points along the card. It is the relation between these pin positions which enables the car to behave appropriately.

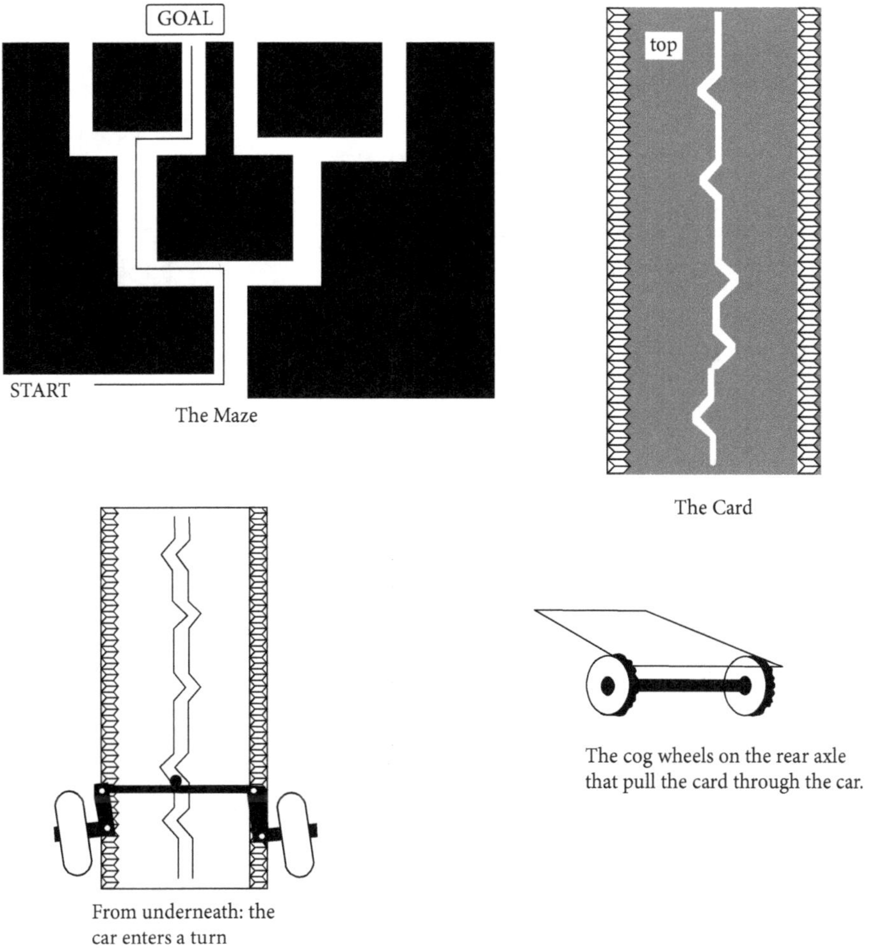

Figure 5.4 Robert Cummins's ingenious case of a driverless car guided by a slot in a card (1996, p. 95).

The pin is driven along the card in a way that correlates with the movement of the vehicle along the track, forwards or backwards, at different speeds. To get a clearer view of the internal processing, imagine it unfolds step-wise, as illustrated in Figure 5.5. The car is not detecting where it is at any moment, so it needs to be started off at a location that correlates with the initial pin position. Suppose this is the start of the track. It then moves forward a certain distance. To work out how it should now align its wheels, it moves the pin forward a corresponding distance in the card and orients the wheels accordingly. This process takes two signals at input, one saying where the system is at the outset, the other correlating with how far it has moved (the turning of the cog wheels). It then makes use of spatial relations between positions on the card to move the pin to the appropriate position on the card, and thus to act appropriately.

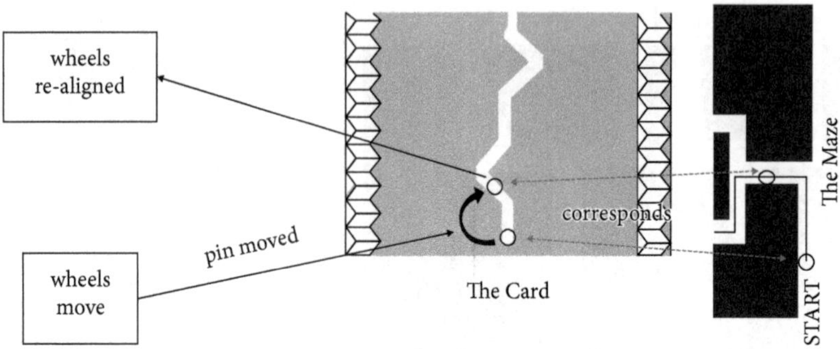

Figure 5.5 One step of the computation being performed in Cummins's driverless car case (detail).

To get to representational content we need to supplement Cummins's case somewhat, so that navigating the track is a task function of the car. We can imagine it has a task function to navigate to the end of the track as a result of robustness plus deliberate design (§3.5). We have robustness if the car is able to get to the end of the track from a range of starting positions, which would be the case if there was a mechanism to ensure that the initial position of the pin in the card correlates with the initial position of the car in the track. These additions leave the basic structure of the case intact. The car then exploits two pieces of correlational information: between the initial position of the pin and starting location; and between the rotation of the cog wheels and the distance moved along the ground. Furthermore, the mechanism exploits the structural correspondence between spatial relations on the card and spatial relations on the track. It is because spatial relations on the card correspond to distance that the system can update the pin position on the card on the basis of information about distance moved (received from the wheels). As a result of this internal computation, lengthways position of the pin on the card remains a correlate of where the car is. Widthways displacement of the pin is an instruction about how to act when at that position. Notice that if lengthways position of the pin were to correlate with location because the car is constantly detecting its current location, rather than just doing so at the outset, the structural correspondence would not be being exploited.

In his influential book *The Organization of Learning*, Randy Gallistel advances a theory of content based on isomorphism. He says he uses representation 'in its mathematical sense', which he glosses as there being what he calls a 'functioning' isomorphism between an aspect of the environment and a brain process that adapts the animal's behaviour to it (Gallistel 1990, pp. 15–33). There is a functioning isomorphism when the correspondence is exploited to solve problems in one domain using operations belonging to the other. This is clearly very like—and indeed partly inspired—my notion of UE structural correspondence. Gallistel has a further requirement: that the isomorphism should be rich, in the sense that there are many operations in the representing domain that correspond to operations in the represented domain. However, in another way his requirement is much weaker than mine.

Gallistel distinguishes between direct and indirect isomorphisms. A direct isomorphism exists where the material or process embodying the representation has properties formally the same as those of the represented material or process (e.g. space mirroring space). There is an indirect isomorphism when there is no formal similarity between the representation and what is represented. For example, a mapping of mass onto written numerical symbols is only an indirect isomorphism, since 'there is no physical ordering of the numerical symbols' (p. 28). Gallistel allows that indirect isomorphisms, where 'the isomorphism is created only by way of an interpretive code', are a sufficient basis for content (p. 28).

That is too liberal, because it would apply to a downstream process that operated something like a look-up table, programming a reaction to each symbol but without relations between the symbols having any significance for the processing. A similarity in the downstream reaction is a kind of relation on the symbols, albeit indirect. (Relations on the downstream outputs other than similarity could also count.) Then there would be an 'indirect isomorphism' on the symbols because of the 'interpretive code' constituted by the downstream reactions. If we allow the interpreter and its dispositions alone to define admissible relations between representations, then we are back to the problem of arbitrary relations between representations counting. We lose the sense of the system making use of an exploitable relation. So Gallistel's indirect isomorphisms will not in general count as cases of exploitable structural correspondence.

However, I do think there is something right in Gallistel's idea that which isomorphisms are relevant is relative to the sensitivity of downstream processing. If processing in the rat hippocampus were not sensitive to the co-activation structure on the place cells, co-activation would not be the basis of an exploitable structural correspondence; if downstream processing then changed so that it became sensitive to relations of co-activation, the structural correspondence would become an exploitable relation. Changes to downstream processing can change which relations on vehicles are being systematically processed, but the relevant relation on vehicles cannot be a relation that exists just in virtue of similarities in the way downstream processing reacts to the vehicles. To be an exploitable structural correspondence, processing must be sensitive in some systematic way to a relation V between vehicles that exists independently of how they are used downstream. Sensitivity here is a causal notion, depending, for example, on the special science laws using projectable predicates that describe the operation of the system. That is important if there is to be a substantive sense in which the structural correspondence is a resource being used by the system. It is not entirely constituted by the way representational vehicles v_i are used.

In short, although exploitable structural correspondence depends upon the sensitivity of downstream processing, it cannot be constituted just by the way downstream processes react to vehicles. So, although exploitable structural correspondence is by no means limited to Gallistel's direct isomorphisms, it is much more limited than Gallistel's category of indirect isomorphism.

5.6 Two More Cases of UE Structural Correspondence

(a) *Similarity structure*

Rat navigation gave us an example of UE structural correspondence (§5.2) and the previous section showed that seemingly obvious cases can fail to qualify. This section examines two more case studies in which a structural correspondence is exploited and is thereby constitutive of content, one involving similarity structure and the other causal structure.[12]

We can define a high-dimensional state space that captures the pattern of firing of a large population of neurons. The firing rate of each neuron in the population defines one axis in the state space. The pattern of activation distributed across the neurons at a time defines a vector in the state space. One measure of how similar two patterns of neural activity are is the distance between the two corresponding vectors in this state space (Figure 5.6). Paul Churchland is the leading proponent in philosophy of the idea that similarity in neural state space is important to the way mental representations function (Churchland 2012, 1998). Recent work analysing the distributed patterns of

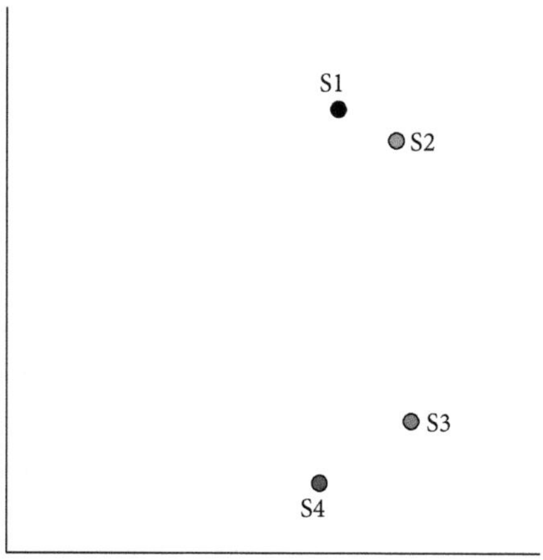

Figure 5.6 Illustration of neural similarity space. The response of two notional neurons to four stimuli S1 to S4. Responses to S1 and S2 are similar to one another and different from S3 and S4. For example, S1 and S2 could be images of faces, S3 and S4 of inanimate objects.

[12] Another obvious case to think about is predication in a natural language sentence. Predication is a relation between representational vehicles (words) and thus is a candidate to form one end of a structural correspondence. Difficulties arise when we ask what relation in the world it corresponds to. Instantiation (of a property by an object) is the obvious candidate, but then the Bradley regress threatens. Since we set aside linguistic representation at the outset, I won't get into those difficulties here.

activation recorded from neurons in non-human animals (Kiani et al. 2007) and recorded by fMRI in humans (Huth et al. 2012) has found cases where the similarity structure of neural activations does indeed mirror the similarity structure of the stimuli presented; for example, of objects of different kinds seen while watching a film.

The existence of a similarity structure does not imply that those similarities are being used computationally, even if the similarities and differences are predictive of some observable effects like differences in reaction times, or repetition suppression in the BOLD response. However, some experiments require subjects to compute similarity; for example, if they are tasked with judging the similarity between various objects. People do so in their own idiosyncratic way. Those judgements are due in some way to how the objects are represented in the brain. Since there is good evidence that the particular structure of an individual's similarity judgements is predicted by the idiosyncratic structure of their neural activation space (Charest et al. 2014), it is likely that similarity between neural activation patterns is the basis on which the individual is making their similarity judgements. That is, subjects are relying on a computation that uses distance in neural activation space as a measure of how similar two objects are. Another experiment used silhouettes of birds that vary along two dimensions (leg length and neck length: Constantinescu et al. 2016). When tasked with morphing an initial silhouette into a given target, subjects revealed that they had grasped the similarity space of the samples on which they had been trained. Again, this corresponded to a neural similarity space that was extracted from patterns in fMRI activation.

In line with these findings, let us suppose that activation space is sometimes used to make similarity judgements. When a subject looks at two images in series, eliciting two distributed patterns of neural activation, a measure is taken of how nearby the two patterns are in activation space. Pairs that are close on this measure are judged to be similar; pairs with a larger neural distance measure are judged to be more dissimilar. Suppose further that subjects have received feedback for correctly judging similarity according to some property of the objects.[13] Sorting objects according to similarity then becomes a stabilized function and, assuming some robustness, thereby a task function.[14] Individual patterns of activation are then being exploited for their correlation with the type of object being viewed; and the relation between two patterns in activation space is being exploited for the fact that it corresponds to the similarity of the objects represented by those two patterns. So, the correspondence between

[13] As in Constantinescu et al. (2016). In that case the dimension of similarity was objective, i.e. not determined by the way people tend to judge or experience the objects' similarities and differences. However, the property could equally be intersubjective, i.e. dependent on how people in general tend to experience the objects (so not fixed by similarity and difference in this subject's individual responses). If the task involves coordinating with others (e.g. in picking a colour scheme), then feedback, hence stabilization, depends on the individual's similarity judgements accurately tracking this intersubjective response-dependent dimension of similarity.

[14] This is a simplification. It would be more realistic to suppose that recognizing objective similarity and difference is a means to performing some different task function.

distance in neural activation space and similarity in the space of objects/properties in the world is a UE structural correspondence.[15]

These experiments raise the issue of the role of subjectively experienced similarity space: similarities and differences in the kind of conscious experience prompted by different images or objects. The experimental findings concern neural similarity space not experiential similarity space; however, a common intuition is that we use experienced similarity when judging the similarity between different objects. That is not the claim made here. My claim that relations between patterns of neural activation can structurally represent similarity between objects does not depend on these similarities and differences being experienced by the subject. The way content arises out of relations between vehicles does not depend on those relations being apparent at the personal level.

(b) Causal structure

The second case involves causal structure. The cognitive details are less clear, but the case is important because the ability to represent causal structure has been so significant for the evolution of human cognition. It is through understanding causal structure that we are able to assess the effects of various interventions. For example, we can observe that a falling barometer needle predicts that it will rain but, understanding the causal structure, we wouldn't try to make it rain by moving the barometer needle. Causal understanding is crucial to human tool use and technology.

Many animals can learn which action is best to perform in a situation. A simple way to do that is to keep track of the consequences of performing each action, and to value an action more when it produces good consequences. That way of learning does not record what the consequences were, just whether they were good or bad. It is called 'model-free' or 'habit-based' learning. It does not involve a causal model of how actions produce their consequences. The animal gets into the habit of performing an action when it has repeatedly led to good consequences. Action A could get a high value because it leads to water and happened to have been performed when thirsty. If the animal is no longer thirsty, then getting water is no longer rewarding, but action A would still be chosen. It takes a number of trials to learn that action A now no longer leads to rewarding consequences. A system with knowledge of causal structure, by contrast, can represent that action A leads to water. This allows a person to calculate, when they are not thirsty, that the consequences of performing action A are no longer valuable. They can refrain from choosing it without having to experience the consequences. Decisions based on reasoning with a causal model of actions and their consequences are called 'model-based' or 'goal-directed' (Dayan 2014). The habitual tendencies

[15] If the neural activation space arises as a result of training, as in neural network models, then this is another case where the exploitable structural correspondence arises at the same time as it is stabilized (§5.4a).

produced by the model-free system can be inhibited to allow the person to choose a model-based or goal-directed response.

A now-classic way to test for model-based reasoning, hence for knowledge of causal structure, is the two-step task (Gläscher et al. 2010). This adds probability into the picture. Suppose you are presented with sweets wrapped in black or white wrappers, one colour for strawberry, the other for lemon, and you don't know which is which. The sweets are in two jars, jar A has mostly black sweets, jar B mostly white. You like lemon and hate strawberry. You reach into jar A, which is mostly black, but happen to get a white sweet, and find that it is lemon-flavoured. Your action, reaching into jar A, was rewarded. So, the model-free system would incline you to do it again. Instead you reason that you are more likely to get the lemon flavour you want from jar B, because white sweets are much more numerous there. So you do the opposite of your previously rewarded action and reach into jar B. Experiments with this logic show that human subjects do select actions based on knowledge of causal structure (Gläscher et al. 2010, Daw et al. 2011). However, we have not yet reached structural representation, because the computations involved in this reasoning only require correlation-based representations of states and of transition probabilities between states (Daw and Dayan 2014).

A more complicated experiment does give us evidence that humans have structural representations of causal structure. Quentin Huys and colleagues trained subjects on the task structure illustrated in Figure 5.7 (Huys et al. 2012, Huys et al. 2015). Think of making a series of left–right choices as you move through a maze. Subjects had to make a series of three to five binary choices to pass between six boxes, with the cost or benefit of each choice dependent on which box the subject was in when choosing. For example, when in box 1 a left button press produces a reward of 140 pence and a right button press a reward of 20 pence. Subjects never see the structure of the task but have to learn it by making a series of choices and getting feedback.[16] Huys et al. were able to test rival models of which calculations were driving subjects' behaviour and obtained good evidence that subjects were indeed evaluating in advance the overall benefit of possible sequences of choices before making their decisions. These calculations involve partial searches and maladaptive 'pruning': subjects overlook optimal sequences if they involve a large initial loss.

Causal planning is likely to depend on representations in the prefrontal cortex, especially when a hierarchy of steps is involved (Koechlin et al. 2003; Passingham 2008, pp. 168–70; Koechlin and Hyafil 2007; Balaguer et al. 2016). Understanding how a series of actions and events are causally linked may be an elaboration of the ability to represent the sequential order of events. We saw above that the rat hippocampus will replay activity corresponding to a sequence of locations the animal has visited.

[16] In many causal learning experiments, subjects have to learn about causal structure during reinforcement, i.e. while they are learning how to behave in reliance on the structural correspondence which is being created, e.g. Goodman et al. (2007). So, these are further cases where the exploitable structural correspondence comes into existence while it is being stabilized (cp. §5.4a).

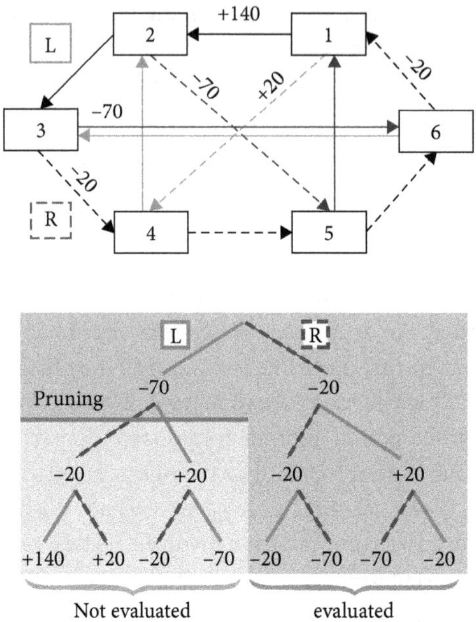

Figure 5.7 The top panel shows the structure of the task studied by Huys et al. (2012, 2015). Arrows starting at each box are labelled with the reward or cost of choosing the left option (solid arrow) or right option (broken arrow) at that step. E.g. choosing right in box 1 produces a small gain of 20 pence and leads to box 4. The bottom panel shows one of the decision trees which subjects think through when they are evaluating possible paths through the structure starting from box 3. Subjects don't evaluate the left-hand branch (solid line) beyond the first step, since it involves incurring a large initial loss (−70 pence), even though it would be the optimal choice (left, right, left for +50 pence in aggregate).

Similarly in a non-spatial task, when human subjects learn sequences of six visual images, brain activity during rest spontaneously revisits the states it was in when viewing the images, capturing the order in which the images were experienced (Kurth-Nelson et al. 2016).[17] When sequential structure mirrors causal structure, that correspondence is exploitable for the purposes of causal reasoning.

The calculations ingeniously uncovered by Huys et al. (2015) clearly depend on subjects representing the relations between the six states, reasoning through sequences of them, and sometimes cutting that reasoning short when they encounter a large loss. There is not a rich neural story of how the step-by-step reasoning occurs, but the findings of Kurth-Nelson et al. (2016) are suggestive. So, let us suppose that subjects have brain states that occur in sequential order; for example, the state for box 1 potentiates the states for box 2 and box 4, each conditional on a different action (left and right, respectively). When a subject calculates that box 5 is accessible within two steps from

[17] In this experiment the repeatable patterns were measured across the whole brain. The hippocampus alone is unlikely to be coding the images directly, but may be coding the position of an image in a sequence: a distributed pattern of firing specific to an object at a location can be decoded from hippocampal activity (Hsieh et al. 2014).

box 1, that calculation makes use of the sequential structure of brain states, and of the fact that it corresponds to causal structure in the world in which she is making her choices. That would then be a case of UE structural correspondence. The sequential order of neural states is being exploited for its correspondence with the relation of causal accessibility between world states. In the absence of detailed understanding of the neural vehicles, this is more of a 'how-possible' case study. It does show how UE structural correspondence could be a suitable resource to form the basis of structural representations of causal structure.

5.7 Some Further Issues

(a) Exploiting structural correspondence cannot be assimilated to exploiting correlation

An objection to basing content on UE structural correspondence runs as follows: any exploitable structural correspondence carries correlational information and in fact it is the correlational information that is playing the content-constituting role. I agree that in very many cases the relation V involved in a UE structural correspondence will also carry correlational information about the relation H it represents. The relation of co-activation between place cells is learnt. It's being instantiated raises the probability that the two corresponding locations are near to one another. Even if a structure is acquired by evolution, and is not subject to learning during the lifetime of an individual organism, there is still often a sense in which it carries correlational information: had the world been different, the structure would have been different. So, the structure being as it is raises the probability that various relations obtain in the world.

However, the fact that a relation V between representations v_i and v_j carries correlational information does not imply that V's carrying information is being exploited, nor further that it is being exploited for carrying information about the obtaining of a relation *between the entities represented by v_i and v_j*. Think about hierarchical processing; for example, Marr's theory of the stages of processing in the visual system (Marr 1982). Activity at one layer in the hierarchy depends on the activities of vehicles lower down the hierarchy, in particular on relations between them. For a simplified example, consider the way angular disparity between the two eyes is used as a depth cue (see Figure 5.8). When the eyes focus on an object, the more their viewing angles converge, the closer the object is. Various signals in the brain correlate with eye gaze direction: let's suppose state A is a firing rate that correlates with and represents the horizontal angle of the left eye, state B of the right. The difference between rate A and rate B correlates inversely with the distance of the object of focal attention. That is, a relation between A and B, call it C, correlates with the distance of the object. Suppose downstream processing makes use of this relation C in a way that depends on the distance of the object; for example, by programming reaching movements that correlate with C. Is C thereby a structural representation?

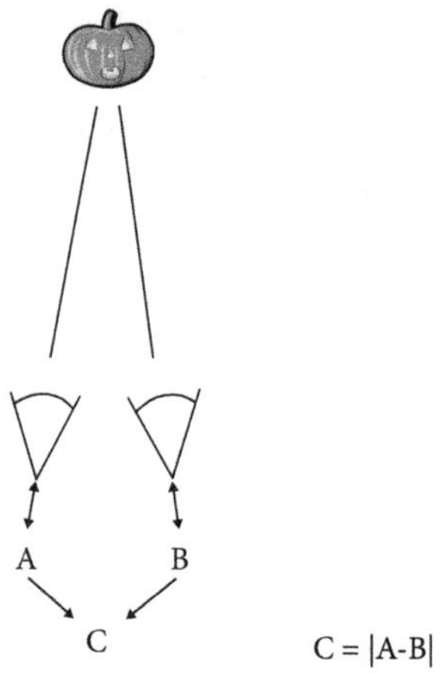

Figure 5.8 Firing rates A and B correlate with the current orientation of the left and right eyes, respectively. The (unsigned) difference in firing rates C correlates inversely with distance to the current object of focal attention.

To qualify as a structural representation, the relation C on vehicles A and B would have to represent a relation on the entities represented by A and B (see definition in §5.2 above). That is not the case here. The content of C is something like *the attended object is at distance x*. A and B represent eye direction (e.g. something like *the left eye is pointing at angle θ*). C is not representing a relation between the entities that figure in the contents carried by A and B. Hierarchical processing will make use of relations between representations to extract further useful information from them. That involves coming to represent a new condition that could be inferred probabilistically from the conditions already represented. It does not generally involve representing a relation between entities already represented.

A different line of objection is that my account of UE information already trades on a second-order resemblance theory of content. I have a collection of internal vehicles performing a computation. The functional relations between vehicles seem to correspond to relations between the entities they represent. For example, vehicles representing local motion and local colour are transformed into a vehicle representing coherent motion (§4.7). That functional transition seems to correspond to a relation in the world: surfaces exhibiting such-and-such local chromatic patterns tend to be moving thus-and-so. Doesn't the whole story about internal vehicles implementing algorithms depend on functional resemblance fixing content; that is, on a second-order correspondence at the level of computational structure (see O'Brien and Opie 2004)? The answer is that a computational step is not itself a structural representation. It does not represent that a relation obtains in the world. The utility of a computational step might depend on the presupposition that p (e.g. that such-and-such chromatic properties are a sign of sameness of surface). We could even say that the system implicitly represents that p (Shea 2015). But this is not a content for which there is a vehicle. The information that p obtains is not available to be calculated with, to be used in other computational steps. You can call this a 'computational structure' if you like, but that does not entail that structural representations are involved.

So, most cases where a relation between representations is exploited for its correlational information, and therefore carries UE information, do not qualify as cases of structural representation. Exploiting structural correspondence is a special kind of case, which makes it worthwhile to pick it out and analyse it separately. And, indeed, the way content is constituted works differently.[18] That has two consequences. The first consequence we saw with place cells: a new place cell would have content in virtue of its place in the co-activation structure, irrespective of any online correlational properties. With structural representations based on UE structural correspondence, since the same relation has a systematic significance across a range of representational vehicles, new representational vehicles that fall under the relation can acquire content in a way that is independent of their correlational properties. The second consequence is exemplified by the way co-activation is used to calculate efficient routes: the relation is available to be used in computations across a range of vehicles in a systematic way. Neither of these features necessarily accompanies UE information.

Furthermore, it is at least conceptually possible for there to be a UE structural correspondence that carries no correlational information at all. An ant crawling in the sand could by chance trace a figure that looks like Winston Churchill (Putnam 1981, p. 1). The sand figure would carry no correlational information, but someone noticing the structural correspondence could use the figure to make calculations (e.g. comparing eye separation to nose length). Similarly, in subpersonal cases, a structure that happened just by chance to correspond in useful ways to significant entities and properties in the world would be useful to an organism, even though the structure's being that way is accidental—it carries no information about relations in the world of significance to the organism. It is not so far-fetched that there could be accidental structural correspondences that neural computations can make use of. Neural activity can organize spontaneously into cycles, automatically proceeding through a repeating series of steps.[19] Such a cycle bears a structural correspondence to all kinds of cyclical processes in the world (recall the liberalism) without carrying information about them. For example, a ten-stage neural cycle corresponds to ten major stages in the life cycle of a perennial plant. Neural processing can readily become sensitive to the time that it takes to transition between states of a rapid neural cycle. Then temporal relations

[18] Karen Neander's recent book makes second-order resemblance constitutive of content in some cases (e.g. for perceptual states); however, she sees second-order resemblance as a supplement to her causal-teleosemantic theory (Neander 2017, pp. 175–215), where I take structural correspondence to be an alternative basis of content. (Also, my notion of structural correspondence is not limited to relations that meet the conditions for being a similarity/distance relation.) Neander's account has the same attractive consequence that it fixes content for new representational vehicles that fall under the same relation.

[19] The repeating hexagonal pattern of grid cells is another candidate (Constantinescu et al. 2016). This structure can be used for its correspondence to relations in the world with the same structure, when that is relevant to a new task, even though that is not why the neural structure exists.

between stages in the plant cycle could be computed by using the (much shorter) temporal relations between states in the neural cycle. In this way a purely accidental correspondence would come to be a UE structural correspondence.

In sum, there are good reasons for a theory of content to pick out UE structural correspondence separately from UE information as a basis for the existence of representational content.

(b) Approximate instantiation

The definition of structural correspondence we have been using is an exact one. That is an idealization. The way that a structural correspondence explains task functions is that instances of it are instantiated[20] on occasions when the task function is stabilized and robustly produced. The correspondence does not need to be exact on those occasions in order for the structural correspondence to be explanatory. (Similarly, a correlation does not need to be perfect for correlational information to be explanatory.) A correspondence with two locations being roughly 10cm apart during stabilization can explain an organism's performance of task functions.

Consider a structural correspondence I that maps co-activation to distance with a certain metric, and consider a certain co-activation delay V that occurs between the activity of two place cells v_i and v_j. I maps these to locations x_i and x_j and maps V to a spatial separation of 1 cm. We can say I is *approximately instantiated* on occasions when the actual distance between x_i and x_j is approximately equal to the distance to which V maps under I, in this case 1 cm. The explanandum is near-optimal behaviour, and the fact that I is approximately instantiated can explain why the rat chooses a near-optimal route that passes through x_i and x_j.

If we didn't include structural correspondences that are approximately instantiated, then the existence of an exploitable structural correspondence would be a very demanding constraint indeed. The definition of exploitable structural correspondence puts strong restrictions on the relations that are candidates on both sides of the correspondence. In the real world it will almost never be the case that there is an exact correspondence between these relations. Requiring that such a tightly restricted correspondence should be exactly instantiated, in order for our theory to have recourse to it, would be an overly demanding constraint.

However, once we allow approximate instantiation we open up a whole class of candidate exploitable structural correspondences. Distance is a relation of significance to the rat, but co-activation maps smoothly to distance in continuum-many ways, placing different metrics on the locations represented. Which of these mappings gives the content? We answer that by looking for relations that play an unmediated role in explaining S's performance of task functions, allowing for approximate instantiation. For each exploitable structural correspondence I, we can ask how approximately or exactly it was instantiated across the range of cases that were involved in stabilizing

[20] Defined at the end of §5.4a above.

the system's task functions and producing them robustly. Suppose I maps $V(v_i,v_j)$ to $H(x_i,x_j)$. We can consider all the occasions when tokening of representational vehicles figures in the explanation of task functions and calculate how closely the actual relation between x_i and x_j matches H across those occasions (e.g. how nearly their actual spatial separation matches the distance H given under *I*).[21] The sum of those values across instantiations (possibly weighted for their significance) measures how accurately or approximately I was instantiated across those occasions.

By repeating this process for all the many candidate correspondences, we get a measure for each. Generally, instantiation being less approximate will make a correspondence a better candidate for being a UE structural correspondence. But just as the content-constituting correlation needn't be the one that maximizes accuracy (Godfrey-Smith 1989), the content-constituting correspondence needn't be the least approximate one. We are in the business of explaining robustness and stabilization, so the degree to which the UE structural correspondence is approximately instantiated during stabilization should match the extent to which episodes of behaviour did eventuate in stabilization-producing feedback. As well as metrical changes, there are also correspondences with different degrees of determinateness. There is a mapping of co-activation to precise distances (e.g. 12.4 cm apart) and another mapping to determinables like *far apart* and *quite close*. Here too we are looking for a degree of determinacy that matches the degree to which instantiation contributed positively to stabilization.[22] These considerations may not settle on a unique candidate, but only arrive at a family of equally explanatory UE structural correspondence relations (e.g. with slightly different metrics), in which case there will be an equivalent degree of indeterminacy in the content.

The degree of approximate instantiation is only a subsidiary consideration in homing in on UE structural correspondence. The primary concern remains finding a correspondence with objects and properties that figure directly in a causal explanation of robustness and stabilization: how robustly produced outcomes had consequences in the world that produced effects on the organism which stabilized this behavioural tendency and the mechanism which produced it. Approximate instantiation comes in when we are deciding between different mappings to explanatory objects and properties, for example different metrics for mapping temporal differences in neural firing onto spatial differences between locations. Suppose, hypothetically, that a mapping from rat place cell co-activation to light intensity differences was more accurately instantiated than the mapping to space when its task functions were stabilized. That mapping would be a less good candidate because light intensity differences could only provide a mediated explanation of spatial route-finding behaviour. Locations, distances, and

[21] In cases considered at the end of §5.4 above, where the referent of the vehicles v_i and v_j is not already fixed (e.g. by UE information), we also need to consider how permutations of their referents would affect accuracy.

[22] Optimality is a special case of this. One approach to representation in cognitive science is to lean heavily on optimality. Organisms are said to represent contents that make them cognitively optimal in some sense (e.g. Bayes rational). From our perspective that is a special case of this more general principle.

rewards at locations figure directly in a causal explanation of how rat navigation behaviour is stabilized. Light intensity could only be explanatory because it correlates with these causally relevant properties.

This way of handling the approximation inherent in occasions when a real organism performs real behaviour can, I think, also handle representational redundancy. The definition of structural correspondence we have been working with follows the mathematical notion of homomorphism. Since the mapping need not be one-to-one, two vehicles may be mapped to the same entity (e.g. v_i and v_j both to x_i).[23] But suppose two place cells map to the same location, and that one activates the other. Co-activation would then represent that they are some small distance apart, which of course cannot be the case if they both map to the same location. So, this would be a case where the relation represented under the mapping (being a small distance apart) is only approximately instantiated on the occasions that go into explaining task functions (where there is no distance between the locations mapped, since they are both the same location). Thus, representational redundancy will increase the extent to which a structural correspondence is only approximately instantiated, but mappings that contain some redundancy are not excluded from being candidates for a UE structural correspondence. Similarly, we can compare approximateness between correspondences under which the mapping of vehicles v_i to worldly entities x_j has been permuted.

(c) *Evidential test for UE structural correspondence*

The idea of approximate instantiation gives us another useful tool. When discussing UE information in the last chapter, I suggested a rough-and-ready evidential test (§4.2). The correlation whose strengthening and weakening is most directly tied to the likelihood of the system achieving its task functions is a good candidate to be UE information. We now have the tools to formulate a similar evidential test for UE structural correspondence. Here we look at how accurately or approximately a correspondence obtains on occasions when it is instantiated. We then apply the same idea. For a candidate structural correspondence I, we see what the effect would be if it were instantiated more accurately. Would the system be more likely to achieve its task functions? A correspondence for which the accuracy of its instantiation is more directly connected to the likelihood of achieving task functions is a better candidate for content.

> *Evidential test for UE structural correspondence*
> The exploitable structural correspondence defined on putative vehicles of content in a system S performing task functions F_j which is such that
>
> > being less approximately instantiated most increases and being more approximately instantiated most decreases the likelihood of S achieving F_j
>
> is a good candidate to be a UE structural correspondence.

[23] I.e. homomorphism allows functions that are not surjective.

As we saw before, this test may be empty, indeterminate, or of little practical use. But it will often home in on content. Something like it is often at work epistemically in assigning content in real cases in cognitive neuroscience. The varitel framework allows us to see why that should be. The test also helps with some of the questions about indeterminacy we saw in the last section. Rat place cells have a less determinate and a more determinate mapping to distance (quite far away vs. 22.4 cm away). The evidential test counts against the less determinate mapping.

As before, the test is only applied to objects and properties in the world that are of significance to the organism, so it is in some ways subsidiary to constraints deriving from causal explanations of task functions. It is important to note that it does not imply that content is given by the most accurate correspondence (the one which is least approximately instantiated). It tests for how much changes in accuracy would impact on the likelihood of S producing task functions F_j and receiving stabilizing feedback. For example, prey animals frequently dart away from noises. The occasions which contribute to stabilization, that is when a predator is present, are much rarer (Godfrey-Smith 1991). However, on those occasions it is the relation with predators that has the most direct effect on whether the hapless prey achieves the task function of avoiding predators.

To see the test in action, let's revisit the experiment performed by Constantinescu et al. (2016). They obtained evidence that subjects had learnt a two-dimensional space for a series of cartoon birds, with the dimensions defined by leg length and neck length. They found that distance N in neural activation space corresponds to similarity S_{2D} in that two-dimensional feature space. Furthermore, this correspondence explains how subjects are able to move from a starting state to a target image with the minimal amount of adjustment. Now consider a different (but closely related) candidate structural correspondence: the correspondence between neural activation distance N and the leg-length dimension taken alone S_{1D}. How accurately or approximately neural distance N mirrors leg-length would also have an impact on the likelihood of the subject achieving an efficient adjustment to reach the target image. But it would have less of an effect on achieving that outcome than the correspondence between N and S_{2D}. Consider another even weaker candidate: the correspondence between N and the overall size of the image. Instantiating that correspondence more accurately when performing the task would have a negligible effect on task performance, it might even decrease it. So, in this case the evidential test plausibly picks out the UE structural correspondence (the 2D feature space).

5.8 Conclusion

Representations are stand-ins. What better stand-in than symbols that are isomorphic to the domain you are reasoning about? It is a small step from that observation to the claim that content is based on isomorphism, homomorphism or structural correspondence. That cannot just be a matter of first order resemblance, but once we cast the net more widely, it is unclear where to stop: the standard objection is that second-order resemblance, isomorphism, and other correspondence relations are too liberal to contribute any substantial restriction to a plausible theory of content. If content were ubiquitous it would lose its explanatory purchase. From our perspective—in which a theory of content is constrained by the explanatory role of representation—this liberality is a symptom of a deeper problem. The vast majority of structural correspondences which exist are not usable. And even where there is an obvious and sometimes exploitable structural correspondence, it is often not being used by the system doing the representing. On the other hand, where a system is systematically sensitive to a relation on a collection of vehicles, having that relation correspond to a relation in the world that matters to the organism—that is significant for the performance of task functions—is a very substantial achievement indeed. In this chapter we saw that instances of such an exploitable structural correspondence can take centre stage in explaining how an organism performs task functions. By this route, structural correspondence is a basis of content: it is a necessary part of a sufficient condition for content determination.

PART III

6

Standard Objections

6.1 Introduction

The positive story is now on the table. We have seen how different accounts of content are suited to dealing with the representations involved in different cases. For each case study, the aim was to give a theory of content that is empirically well-supported and accounts for the way contents are used to explain behaviour. The varitel framework was also designed to produce accounts of content which overcome the most important objections to teleosemantic and other existing theories of content. This chapter will address those challenges explicitly, referring to the existing literature in more detail than when setting out the positive accounts above.

Section 6.2 shows how the approach deals with problems of indeterminacy: distality, disjunction, the qua problem, and so on. The accounts do not deliver perfectly determinate contents, but I will argue that the level of determinacy achieved is appropriate to the nature of the systems whose behaviour is being explained. Section 6.3 turns to systematicity and productivity, pointing out that the systems we have been discussing do not generally show the kind of compositionality present in natural language sentences. In §6.4 we look at swampman and related challenges to the idea that representational content should depend on a system's history. Finally, in §6.5 we briefly ask what kind of normativity attaches to representational contents of the kind we have been discussing here and consider the objection that the normativity of representation is categorically different from, and so cannot be based on, the normativity of function (which has not been my aim).

6.2 Indeterminacy

(a) *Aspects of the problem*

A few examples are standardly used to pose indeterminacy problems for theories of content, most prominently the frog's tongue-dart reflex. I will use that example to illustrate the way my approach deals with various aspects of indeterminacy. Then I will raise and answer those problems for two of our case studies: analogue magnitude representations (in §6.2(b) and (c) below) and cognitive maps (in §6.2(d) below).

I follow the literature in simplifying the frog case. This stylized treatment serves to illustrate the key philosophical issues. So, let us suppose that the frog's tongue dart is triggered by the activity of an array of neurons in the retinal ganglion. Each neuron triggers a tongue dart to a particular location when its activity crosses a threshold. There is mutual inhibition so that only one cell crosses the threshold at any one time.

Consider the putative representation R constituted by the firing of the cell which triggers a tongue dart to location (x, y, z). In typical cases this is caused by a passing fly. Light reflected off the fly and its surroundings passes through the air, through the frog's eye and hits the retina. The pattern of light and shadow hitting the retina excites the retinal ganglion cell R. This causes the tongue to dart out towards the fly at (x, y, z), which is trapped and ingested by the frog. Nutrients from the fly contribute to the frog's survival, and may thereby contribute to the number of offspring it produces. The tongue-dart response is specific to stimuli with tightly delineated characteristics, nevertheless the sensitivity of the system is such that the frog will also snap at little black things that are not flies, like a moving pellet on the end of a fine wire.

The distality problem is to specify which stage of this causal chain contents are concerned with: at the fly or other passing object, or further along the causal chain to the frog's retina and retinal ganglion. At the other end, content could concern proximal effects like the firing of motor neurons or the movement of the tongue, or more distal effects like catching the fly, digesting its nutrients, or eventual continued survival and reproduction.

The specificity problem arises for a given stage in this causal chain. For example, a passing fly has many properties. It is a little black thing, a fly (biological taxon), a flying nutritious object (ecological category), something worth eating, something good for the frog, and something that will promote reproductive fitness. R's being tokened increases the probability that every one of these conditions obtains at location (x,y,z). They are not coextensional:[1] R can be set off by little black things that are not flies. Another aspect of specificity arises particularly with teleosemantics, because of its reliance on conditions under which behaviour prompted by R promoted survival and reproduction. That seems to let all sorts of non-specific conditions into the picture: that the prey is not poisonous, that there is no predator nearby that will be alerted to the frog's presence; also background conditions like the presence of air between frog and prey, and normal gravitational forces.

Finally, there is a disjunction problem in that any two or more of these conditions can be put together to produce another condition with which R also correlates. For example, R correlates with: *flying nutritious object at (x,y,z) or little black pellet at (x,y,z)*. In its specific form the disjunction problem is the problem that, for any two worldly conditions C1 and C2 which are candidates for the content of R, C1-or-C2 is a candidate. If R carries correlational information about each condition it will more strongly probabilify their disjunction: $P(C1vC2|R) \geq P(C1|R)$. 'Disjunction problem' is often used more broadly as an umbrella term for all these kinds of indeterminacy. I call them all instances of the problem of determinacy or indeterminacy.

In the past, discussions of indeterminacy have got bogged down in swapping intuitions about what is represented in a given case. Where the representations in question are things like beliefs, desires, and conscious states, we at least have some reason to think that our intuitions about content could get some traction. With the firing in the frog's retinal ganglion cells, we have no such reassurance. There is little reason to give any weight to intuitive judgements about what is represented. The same applies to the case studies we have examined. Instead we have been asking how representations explain behaviour, and what representational contents can underpin those explanations. In Chapter 2 I argued that these explanatory practices are an appropriate constraint on theorizing about content. So, the test of a theory is not that it should deliver intuitive content attributions, but that it should deliver content attributions suited for the explanations of behaviour in which representations figure. Whether or not the contents are appropriately determinate needs to be assessed in that light.

[1] Success conditions and satisfaction conditions generally involve the instantiation of a property by a particular. Since we are leaving aside neo-Fregean sense, it does not matter how those properties and particulars are picked out. Different descriptions can pick out the same success condition, e.g. *fly at (x,y,z)* or *fly at my favourite location* pick out the same success condition (assuming I particularly like location (x,y,z) for some reason). Where the property picked out differs, the success conditions are different, even if those properties happen to have the same extension, e.g. *renate animal at (x,y,z)* ≠ *chordate animal at (x,y,z)*.

(b) Determinacy of task functions

Our first resource is the determinacy of task functions. Recall that task functions are robust outcomes that have a stabilized function or a design function. Out of all the factors that do or could affect the internal processing that produces an outcome F or the consequences that flow from producing F, only a few would be cited in a causal explanation of how F was produced and systematically stabilized by natural selection, learning, or contributing to the persistence of the organism (§3.4d).[2] That is a very substantial restriction on candidate contents.

To apply my framework to the case of the frog, we need to identify task functions and the internal mechanisms that subserve them. The fly-capture mechanism gives the frog a disposition to catch passing flies, which is robust in the face of perturbations and different starting positions. That is, a robust outcome function of the behaviour prompted by R is to trap a fly at (x, y, z). Plausibly, this disposition is the result of selection: past organisms achieving the output of capturing flies is part of the explanation of why there are systems around today with this robust disposition. So catching a fly at (x, y, z) is a task function of R. This task function is achieved synchronically by having an internal mechanism that gathers incoming information, activates one of a range of possible intermediate states (R), and executes a tongue strike in a corresponding direction. That is a simple algorithm, which makes use of the correlation between R and the location of flies on the input side, and between R and the direction of a tongue dart on the output side.[3]

There are a variety of ways of describing the outcome that contributed to selection; for example: catching a fly, a flying nutritious object, or a little black thing. Fodor has argued that, if the categories *fly* and *little black thing* were coextensional in the history of the frog, then considerations based on natural selection cannot choose between them (Fodor 1990, p. 72). That is mistaken. Selection is a causal process. Causal explanation does not in general permit substitution of coextensional properties. Facts about what has been selected for are based on these causal explanations (Godfrey-Smith 1994a, p. 273; 2008)[4]. Only some of the downstream effects of tokening R have been responsible for that disposition's contribution to survival and reproduction. Properties like being small and black do not cause the frog to survive or reproduce (cp. Price 2001, ch. 5, §2). It is because something nutritious was captured that the behavioural disposition was selected. This excludes *little black thing*. It also excludes *something that will promote fitness*, since that is a restatement rather than an explanation of how an outcome leads to survival and reproduction.

Considerations about selection/stabilization are sometimes thought to generate very detailed contents: R represents that there is a flying nutritious object at (x, y, z)

[2] I leave aside design function in the following discussion, but appealing to the intentions of a deliberate designer is obviously another way to secure determinacy.

[3] Without computations over internal components, this case has the same structure as animal signalling cases like the honeybee nectar dance (Chapter 1), which are more straightforward for consumer-based teleosemantic views to deal with.

[4] Sober (1994) makes the distinction between selection of and selection for.

that is not poisonous, contains proteins needed by the frog's physiology, is not moving too fast to be caught, etc. These conditions are of course relevant to whether behaviour prompted by R was stabilized. However, it is a general feature of causal explanations that they do not mention all potentially relevant details, still less the absence of potential defeaters. Explanation has something to do with capturing patterns and generalizing across many events. That counts against picking out events in very fine-grained ways. This is not the place for a general theory of causal explanation, so I rest with the observation that task functions inherit the determinacy of causal explanations of stabilization. That also counts against background conditions figuring in the content, conditions like *fly at (x,y,z) and gravitation is normal*, or *fly at (x,y,z) in air or some similar light-transmission medium*.

This deals with some specificity issues but leaves others open. It does not choose between the following as the type of object to be captured at (x,y,z): fly, flying nutritious object, or object worth eating. We need to look at how task functions converge with correlational information to choose between these contents, at least between some of them (see next subsection).

Let's see how the determinacy of task functions helps in one of our case studies of UE information (Chapter 4). Consider the analogue magnitude system. It is deployed in situations where behaviour is conditioned on the relative numerosity of collections of objects, doing so by using internal correlates of numerosity and comparing them. We can make the plausible assumption that behaviour involving this system results from reward-based learning, where the rewards arise from acting based on the numerosity of the objects tracked. For example, the animal or person has had the chance to learn that, across a range of different rewarding collections of objects, selecting the more numerous collection leads to higher reward. In animal experiments the reward schedule is devised by the experimenter so we can be confident that numerosity is the basis of reward. It is plausible that some natural learning situations also have that structure, and that learning of this kind, aversive as well as appetitive, underpins behaviour driven by the analogue magnitude system. If we simplify to consider just these comparative situations, the analogue magnitude system is an intermediate in achieving the task function of selecting the more numerous collection of objects. That goes a considerable way to making *numerosity*, rather than other related properties, figure in the contents represented.

(c) Correlations that play an unmediated role in explaining task functions

So, determinacy of content flows in part from task functions not being unduly indeterminate. Further determinacy derives from the requirement that the correlations that are content-constituting are those which play an unmediated role in explaining the system's ability to achieve its task functions. This calls for a convergence between correlations and task functions. As a result, the appropriate level of distality is constrained by the task function being explained. With the frog, that place is at the location of the fly. The correlation between R and the location of a fly offers an unmediated explanation of the ability to capture flies at (x,y,z). The correlation between R and a pattern S of light and shadow on the retina could be used to explain fly capture, but less directly: because S correlates with there being a fly at (x,y,z), R's correlation with S can help explain how the frog manages to catch flies. Even though the correlation between R and S may be tighter, this is a mediated explanation. At the input it is the correlation between R and something distal, at the location of the fly, which plays an unmediated role in explaining the achievement of the task function. At the output end, R correlates with producing the distal result (capturing a fly) which is the task function. When asking why the whole R-involving mechanism was stabilized, the fact that R correlates with catching flies figures unmediatedly in the explanation. The same considerations imply that R does not end up representing all the intermediate links in the causal chain from fly to R to fly capture.

The same reasoning applies to the analogue magnitude system. It is activated on the basis of prior internal states that track individual objects (or events or other entities). So, its states correlate with prior internal states, as well as with patterns of light on the retina, or sound in the ear, etc.; also with causal intermediates between the array of objects and the organism. However, the correlations which directly explain how the organism achieves the task function of selecting the more numerous collection are correlations with a property (numerosity) of the collections being selected amongst.

Some indeterminacies left open by considering task functions and causal explanations of stabilization are resolved when we ask how a collection of correlations carried by a collection of components explains how task functions are performed. We just saw that causal explanations of stabilization might not choose between *fly at (x,y,z)* and *object worth eating at (x,y,z)*. But the frog's fly-capture tongue-dart mechanism is just one of the ways it gets prey. Other internal states correlate with other types of object worth eating to allow the frogs to ingest those. Saying they all just represent *object worth eating* would not capture relevant differences. So, the correlation with flies offers a more perspicuous explanation of how the whole organism achieves its suite of task functions: different mechanisms subserve different tasks and do so in virtue of different correlations.

Varitel semantics does not require that the organism should be able to discriminate the conditions it represents. The frog cannot distinguish flies from little moving black things. Nevertheless, R's UE information concerns flies but not little moving black things. R correlates with flies by being sensitive to sensory features that are a good-enough but imperfect sign of flies, but exploitable correlational information is not restricted to the conditions that a vehicle is the most sensitive or specific sign of. Neither the definition of UE information (§4.2a), nor our evidential test (§4.2c), imply that a stronger correlation trumps a weaker one in constituting content.[5]

[5] We noted in Chapter 3 that cognitive scientists often take strong correlation to be an indication of what a vehicle represents, but that is because strong correlation is an indication of what a system has evolved or

There has to be convergence between correlation and stabilization, which means contents need to concern conditions that can figure both in causal explanations of stabilization and nomologically based correlations. Perhaps the tongue-dart behaviour in some species of frog has in fact been stabilized in evolution because of trapping just three different fly species S1, S2, and S3 that are prominent in its ecological setting. Then the behaviour was stabilized by catching S1s at (x, y, z), and by catching S2s and (x, y, z), and by catching S3s at (x, y, z). Turning to the correlations carried by R, however, a disjunctive category like S1-or-S2-or-S3 is unlikely to figure in UE information since disjunctive properties are generally poor candidates to figure in causal explanations.[6] The non-disjunctive category fly (the biological taxon), or the ecological category flying nutritious object, look to be better candidates to figure in nomologically based generalizations about what correlates with what. Thus, the need for convergence homes in on more determinate contents than task functions alone would.

Some indeterminacy remains. The biological taxonomic category fly and the ecological category flying nutritious object look to be equally good candidates, both to figure in causal explanations of stabilization, and to figure in the causal underpinnings of exploitable correlational information. So the content of R will be indeterminate between *fly at (x,y,z)* and *flying nutritious object at (x,y,z)*. Furthermore, there is some indeterminacy about the biological taxon *fly*. Is the category restricted to insects (e.g. the order *Diptera*) or should it include other flying invertebrates? Should the biological taxon be understood cladistically (i.e. in terms of shared descent) or in some other way (e.g. in terms of shared phenotypic features or shared DNA)? The content of R is likely to be indeterminate between these options. If we use the term 'flyish' loosely for flying insects, flying invertebrates and flying nutritious objects, then we can say that R represents *something flyish at (x,y,z)*—with the caveat that contents in all of our case studies will be somewhat less determinate than suggested by the very precise tools (i.e. words) we use to express them.

Greater determinacy is achieved in the analogue magnitude case because the mechanism has been stabilized in a wider range of situations. Looking for correlations that are explanatory of comparative choice behaviour across a range of different objects homes in on the correlation with numerosity. That is what the system represents.

learnt to track. Strength of correlation is not a way of deciding between contents that are equally good from the point of view of explaining stabilization. The evidential test in §4.2c concerns not the strongest correlation, but the correlation changes in whose strength have the greatest impact on achieving task function performance.

[6] It might also fail to qualify as exploitable correlational information, through lack of a univocal reason. Does R carry exploitable correlational information about there being an object b which is S1 or S2 or S3? That requires there to be regions such that P(S1-or-S2-or-S3(b)|R) > P(S1-or-S2-or-S3(b)) for a univocal reason. The property being disjunctive counts against there being a univocal reason. Suppose S1-or-S2-or-S3 forms a proper subset of the category *being a fly* and P(fly(b)|R) > P(fly(b)) for a univocal reason. That same reason is unlikely to connect to the disjunctive category, except because S1 is a kind of fly and S2 is a kind of fly and S3 is a kind of fly. Mentioning this additional factor would make the reason underlying probability-raising for the disjunctive category non-univocal.

Analogue magnitude states correlate somewhat with other features, like total quantity or total surface area of an array of objects, but it is the correlation with numerosity that explains their common role across a range of contexts (as tested in many ingenious experiments). An accumulator system which operates synchronically just like the analogue magnitude system could be present in simpler organisms, and deployed by them in naturally selected behaviours whose acquisition does not depend on learning. If so, those behaviours could have been selected for more specific functions, e.g. to follow the more numerous shoal of fish. If so, the task function serviced would concern something more specific, like the number of conspecifics, rather than numerosity in general. (There are even simpler accumulator systems that do not depend on prior individuation of objects, and simply reflect mass or quantity. Their functions concern quantities but not numerosity.)

Notice that the account does not rely on a representation being caused by what it represents, for example there being a causal connection between the fly and R. It depends only on R carrying correlational information. Suppose R were activated, not by flies directly, but by patches of light on the ground, and that when a patch of light appears, a prey item is likely to land there a short time later. Then P(prey at (x, y, z) | R) would be high, but prey would play no causal role in the tokening of R. The framework would still entail that R represents the location of prey.

This example has the same structure as Paul Pietroski's case of the snorfs and kimus (Pietroski 1992). Kimus are imaginary creatures that are attracted to the red colour of the sun, causing them to climb hills at dusk, thus avoiding their predators the snorfs, who hunt only in the valleys. Pietroski invites the intuition that kimus must be representing redness, rather than something like *snorf-free zone this way*. He argues that the creatures cannot be representing anything about snorfs, since they have no causal sensitivity to snorfs. (The only kimus that have causally interacted with snorfs are historical ones—the kimus that were eaten, hence selected against.) In my view we should give little weight to intuitions about these cases. In any event, our intuitions doubtless draw on imagining a richer picture in which the kimus have conscious sensory experiences and see redness. Once we drop that, the case is wide open. Given my approach to content, all the correlations which R enters into are candidates for content, irrespective of the causal route to tokening R. Correspondingly, if Pietroski's kimus are as simple as the systems in our case studies, they would end up representing the snorf-free direction, even though they have no causal sensitivity to snorfs.

In short, the need for convergence between exploitable correlational information and causal explanations of stabilization is a source of considerable determinacy.

(d) UE structural correspondence

Turning to UE structural correspondence (Chapter 5), the determinacy issues are similar and are answered in a similar way. The cognitive map in the rat hippocampus represents spatial relations between locations. We relied on UE information carried by place cells to explain route planning, so the convergence we have just discussed between correlation and task function is at work there. Distal correlations with locations figure in an unmediated explanation of task function performance, whereas correlations with sensory features would only offer an indirect explanation. Location is somewhat indeterminate, however, and there is matching indeterminacy in the structural correspondences in play. The co-activation structure on an array of place cells corresponds to absolute locations and the absolute distances between them; to absolute locations and relative distances; to locations picked out relative to some landmarks and their absolute or relative distances; and to locations picked out relative to one another and their absolute or relative distances. There may be general metaphysical reasons why some of these are preferred in a causal explanation of task performance, but that only goes so far. If multiple location- and distance-related features are good candidates for causal explanation in general, then our theory will generate contents that are indeterminate between them.

There is a further, more subtle distinction that we can make with linguistic representations, which may or may not arise with our simpler representations. Place cells act like singular terms, picking out particulars. They could do so indexically, like 'this', 'that', 'here', and 'now' in natural language, or non-indexically like 'London' or '2°W, 10°S'. Individual place cells are clearly saying something more than *I am here now*, since they are reused offline with the same content when calculating shortest routes. But we can ask whether the array of place cells is picking out an array of locations indexically, as something like the locations around here now, or non-indexically, with singular terms that work like names for locations. I can think of three possible answers. The first is that there is a general answer about all these kinds of simple systems; for example, that none of the representations is indexical (or conceivably that they all are). The second answer is that cognitive maps represent in a way that is indeterminate between indexical and non-indexical representational contents. Or thirdly, it may be that the question itself is ill-posed, when asked about a system that does not support a distinction between different ways of picking out its referents. I remain neutral between these answers, accepting that this may be a source of indeterminacy in our case studies.

(e) Natural properties

Since content in these cases is fixed by reference to causal explanations, natural properties will be better candidates. This makes some disjunctive properties unsuited to figure in the content. Arbitrary disjunctions are not good candidates to feature in causal explanations.

This consideration also resists the objection based on 'reduced content' (Peacocke 1992, pp. 129–32). R correlates with there being a fly at-(x, y, z)-and-within-the-organism's-lightcone. That condition certainly applied on all occasions when ancestor frogs interacted with flies in their selection history. However, causal explanations do not in general appeal to these kinds of 'reduced' properties. To give a general characterization of the kinds of properties that are candidates to figure in causal explanations would take us beyond the scope of the present enquiry. It suffices to note here that it is facts about causal explanation that rule out reduced contents.

These points only apply to the kinds of simple systems we are considering here. It is clear that more esoteric contents are not ruled out in more sophisticated representational systems, like human conceptual representation. We persons can represent proximal properties as well as distal properties, contents like *fly and within my lightcone*, and disjunctive contents. Those abilities depend upon the greater complexity of our representational apparatus, especially the combinatorial power of concepts.

(f) Different contents for different vehicles

A final factor at work here is the soft constraint that different representational vehicles should have different contents. That is not an explicit part of what it takes for a correlation to amount to UE information, but it follows in many cases. UE information focuses in on correlational information that is exploited in order to perform a task function. Different vehicles have different effects on downstream processing, so ascribing the same contents to a whole range of different vehicles could miss out on important aspects of the way the system performs task functions, hence would be less explanatory.

For example, suppose we treated all retinal ganglion cells in the frog as having the same content. They carry the information that there is a fly somewhere nearby and trigger catching behaviour. Getting a fly is arguably a task function of all the tongue-dart responses. Retinal ganglion cells carrying information about flies does help to explain how that outcome is achieved. But there are also more specific task functions that go with more specific responses: the function of catching a fly at (x, y, z) is a task function of the response prompted by a particular ganglion cell R. The correlation of R with the coarse-grained condition *there is a fly nearby* could be partly explanatory of achieving that function, but the correlation with *there is a fly at (x,y,z)* is more explanatory. So, the latter gives the content.

Millikan has a similar requirement. Built into her idea of 'most proximate Normal explanation' and 'derived adapted proper function' is the idea that different representations, acted on differently by the consumer, should have different contents (Millikan 1984, pp. 44–5, 97). In my case the requirement is not that every representation within an organism should have a different content. But when there is a stage of processing that admits of a range of mutually incompatible vehicles whose differences make a difference to downstream processing, an explanation of how that processing contributes to performing task functions will generally point to correlational information that is different across those vehicles.[7]

[7] In many cases that is because a range of mutually incompatible vehicles will carry correlational information about a range of states in the sense defined in §4.1a.

> *Soft Constraint: Different Contents for Different Vehicles*
> When a stage of processing can adopt a range of mutually incompatible states R_i, each affecting downstream processing in a different way, correlational information which is different for each of the R_i will generally be a better candidate to be UE information, other things being equal.

In the frog case, it follows that the different retinal ganglion cells represent flies at different locations, rather than all simply representing something like *fly nearby*. In the analogue magnitude case it follows that numerosity is being represented rather than something more coarse-grained like *many* and *few*.

The soft constraint applies to mutually incompatible representational vehicles. There is also the question of whether different components within an overall computational process, elements that can be tokened at the same time, can carry the same content. That does arise, for example the visual system contains multiple representations of the location of an observed object. The soft constraint does not rule out such cases. Nevertheless, if we want to see how internal processing carries out computations that are suited to performing task functions, that will generally require different steps to carry different contents.[8] So there are general explanatory reasons that somewhat count against different elements carrying the same content, without ruling it out in a suitably articulated system.

(g) *The appropriate amount of determinacy*

A final consideration is to ask what the appropriate amount of determinacy is. In these simple cases quite a high degree of indeterminacy may be expected. Lacking many of the moving components of richer representational systems like those found in human belief-desire psychology, it should be no surprise that lower-level systems have more indeterminate contents. In systems with more components those components will often be playing more specialized roles.

In giving representational explanations we are appealing to relational properties of component parts in order to explain the system's behaviour. Components will often stand in a family of closely related relations to a family of closely related distal properties. In the frog, these include the taxonomic biological category *flying insect* and the more physiological category *flying nutritious object*. There is no reason to expect this simple system to support a distinction between representing *flying insects* and *flying nutritious objects*. That is a kind of indeterminacy that flows from the limited complexity of the system.

How best to capture this indeterminacy? One approach is to say that the system carries each of these closely related rival contents, and that we can appeal to any of them in explaining its behaviour. Alternatively, it could be that there is a single natural

[8] Cp. the argument in §4.7 that the UE information in the plaid motion system homes in on different correlations for different components.

property in the vicinity of both candidates that figures in the content, but that we are unable to pick it out exactly, because the language we use is unsuited for doing so, being too precise.[9] On the second option content is not strictly indeterminate, but it consists of a determinate success condition that can only be picked out approximately or disjunctively using the tools of natural language. I don't propose to arbitrate between these options. I rest with the claim that the indeterminacy that remains at this level is unobjectionable.

We noted that less indeterminacy is likely to arise in systems with multiple interacting components. There is also a distinction to be made between indeterminacy at the level of an individual vehicle and indeterminacy at the level of the whole system. This is best illustrated with an example. Recall the system in the prefrontal cortex for deciding the preponderant direction of motion of visual stimuli in one context and the preponderant colour in another context. We saw in §4.6b that applying varitel semantics to this system leaves some residual indeterminacy. The content of the input representation for colour, R_1, is indeterminate between (a) *the majority of dots are red*, and (b) *the colour density is predominantly red*. There is corresponding indeterminacy in the representation C_1 that registers context: between (a) *reward will be based on the colour of the majority of dots on the screen*, and (b) *reward will be based on the predominant colour density on the screen*. However, to be explanatory, a correlation carried by R_1 needs to go with a correlation carried by C_1: (a) with (a) or (b) with (b). There is one set of exploitable correlational information carried by the whole collection of components which includes the two (a) clauses as UE information. There is a second set which includes the two (b) clauses. A disjunctive assignment of (a)-or-(b) to R_1 at the same time as (a)-or-(b) to C_1 will not be UE information. Putting R_1's registering (a) together with C_1's registering (b) is a poor explanation of why the system makes the choice it does. In any event, disjunctive conditions are poor candidates to be exploitable correlational information in the first place (§6.2e above).

So, there are indeterminacies about the overall UE information carried by a system that are not simply recapitulated, component-by-component. Furthermore, the need for UE information to align between components, so that interactions between components make sense in the light of their contents, is a significant constraint on indeterminacy in systems with multiple interacting components. These are both reasons why the residual indeterminacy implied by varitel semantics varies with the complexity of the system in question. That is an appropriate result.

(h) Comparison to other theories

My approach to indeterminacy adopts many of the elements relied on by Millikan's teleosemantics (Millikan 1984, 1989, 1990, 1995, 2004). Contents for Millikan derive from the 'most proximal Normal explanation' of how behaviour prompted by a

[9] Naturalness considerations make that unlikely as between flying insect and flying nutritious object, but the point may apply in other cases, e.g. for locations.

representation led to survival and reproduction. Directive content is the output specific to a representation that features in such an explanation. Descriptive content is the condition, specific to a representation, which explains how those outputs led systematically to survival and reproduction. My own focus is on unmediated explanation of the performance and stabilization of task functions. This may cover a wider range of systems, but retains the merits of Millikan's view: indeterminacy is constrained since causal explanation does not generally allow substitution of coextensional properties *salva veritate*; and also by setting aside mediated causal explanations of stabilization. This makes non-natural or disjunctive properties poor candidates for content for Millikan (1990, p. 334), but as with my account, indeterminacies remain as between properties that have equivalent causal-explanatory significance (Godfrey-Smith 1994a, p. 274).

My requirement for convergence between correlational information carried and task function performed is an additional source of constraint (§6.2c and Shea 2007b, cf. Millikan 2009). I am also perhaps more explicit about the requirement that different representations in the same range should have different contents, and about why that is so (§6.2f). Since I do not attempt to apply my account to conceptual representations or conscious states, I have an argument that the indeterminacies which remain are an attractive feature of the account, rather than a failing (§6.2g). Furthermore, as we saw in Chapter 4, giving up the consumer requirement allows us to deal with systems with multiple interacting components—which thereby have relatively determinate contents—more easily.

Papineau also advances a consumer-based teleosemantic theory. For him the theory applies in the first instance to belief-desire psychology. He argues that desires have determinate contents and act as consumer systems for beliefs, which inherit that determinacy. He used to think that, outside the belief-desire system, teleosemantics results in considerable indeterminacy because multiple systems are equally good candidates to count as consumers (Papineau 2003). He now thinks an idea of Neander's can solve that problem (Neander 1995). A component in a system will indeed have many different nested functions (derived from evolution and/or learning), but teleosemantics should only appeal to its specific function, outputs that it produces on its own in the lowest level description in which it appears as an unanalysed part. This leads to a view in which malfunctions only arise from the failings of the component itself, not from interactions with other components (Papineau 2016). My view goes in a somewhat different direction here, as we will see in a moment.

I follow Price in thinking that the way representational properties feature in the explanation of behaviour should help us to characterize their nature (Price 2001, ch. 4, cp. my desideratum §2.2), also in requiring representations to carry correlational information (for me, in one class of cases). Price adopts Neander's useful distinction between 'high church' and 'low church' teleosemantics (Neander 1995). High church teleosemantics ties content to explanation of the success of behaviour prompted by a representation. Low church teleosemantics focuses on the way representations are produced and ties content to the actual discriminative capacities of the organism. Pietroski's argument about the snorfs and the kimus was a push in the low church direction. Millikan and Papineau argue for the high church view. Dretske (1988) and Ryder (2004) are also in the high church, since they tie content to properties that explain successful behaviour.

Price herself adopts a high church view. She argues that teleological considerations, supplemented with some plausible principles, can deliver determinate contents (2001, ch. 3). Price's immediacy and abstractness conditions have a similar effect to my focus on correlations that enter into an unmediated explanation of how the system performs its task functions. Like Papineau, Price relies on Neander's idea that the relevant functions of a device are things that it can do by itself (in servicing a wider mechanism, or 'governor'). My approach has a rough analogue of this when there are multiple components since I ask what each component contributes to an algorithm realized in the system so as to perform its task functions (Chapter 4). However, my task functions are decidedly not limited to outputs for which a single component is responsible. They are outputs of the whole organism and depend on interactions amongst its components. Nor do I think a vehicle is only misrepresenting when something goes wrong with the component responsible for producing that vehicle. Many cases of misrepresentation are caused by malfunction in upstream components; and even when all the internal processing is operating as it was designed to, misrepresentation can occur when the environment is uncooperative (i.e. unlike it was during stabilization). (I also differ from Price in giving up on the need for a consumer, and in the pluralism that allows for different kinds of exploitable relations and different kinds of functions.)

Karen Neander is the leading proponent of low church teleosemantics (Neander 1995, 2006, 2017). She argues that content concerns the objects and properties an organism is causally sensitive to and should be tied to conditions that it can discriminate between. One argument is based on the idea that a component has not itself malfunctioned if the external environment is uncooperative (Neander 1995). So, for example, if a frog snaps at a little black thing that's not a fly, that should not be counted an error, because there is no malfunction within the detection mechanism.[10] But I have argued that facts about how components of a system interact with one another are not enough to get content explanation off the ground (§2.3). We need to look at how they are designed to interact with the distal environment. Long-armed functions can go wrong when the environment is uncooperative without that being attributable to the failure of any of the internal workings.

A second argument is based on 'response functions' and a detailed case study of the science of prey capture in the toad (Neander 2006, 2017). Neander rightly observes that scientists have been concerned to discover how the toad manages to track prey. That is a different explanandum (2017, p. 119). I link content to explanation of behaviour. The scientists are trying to work out how the toad manages to track—I say represent—prey in its environment accurately enough to survive (2017, p. 108). I don't see why long-armed etiological functions need be tied to discriminative capacities, and I certainly

[10] Note that varitel semantics does not equate misrepresentation with malfunction (see §6.5).

reject the verificationist claim that an organism with non-conceptual representations can only represent what it is capable of discriminating (2017, p. 120). We might well be interested in how an organism manages to discriminate the things that it represents, but to formulate that question we need to leave room for a gap between the things it represents and the way it discriminates those things. Verificationist contents are poor at explaining unsuccessful behaviour—for example, explaining why things go badly for a toad when it moves into an environment rich in little black moving things that are not flies. Basing content on discriminative capacities also means that Neander has to add a special purpose principle to make contents come out as distal (2017, p. 222).

Does my account entail, then, that organisms will never represent perceptual features like being a little black thing and will only ever represent properties like *fly*? Surely the human perceptual system represents features of an object, like its size, shape and velocity, on the way to categorizing it as a *fly*? My account agrees. A suitably articulated system does end up representing more sensorily specific features of an object on the way to representing it under more general categories. We saw that for the visual system in §4.7, where my account delivered representations of chromatic properties and local motion properties. That flows from applying the varitel framework to a system in which information-processing is broken down into multiple interacting components; especially when, as in the human perceptual system, a single perceptual representation feeds into many different kinds of downstream processing and behaviour. So, on my view contents can concern perceptual features of objects, and perceptual systems in complex organisms will typically represent features which they then use to track behaviourally significant categories of object. None of that is found in the toad's simple prey-capture mechanism, at least in the stylized version described here.

Other authors have different proposals about which properties are good candidates for representational content. Ryder works this out with respect to a particular mechanism, SINBAD, whose function is to detect statistical regularities in patterns of input (Ryder 2004). As a result, SINBAD's states end up referring to properties that explain those regularities. Martínez makes an ontologically more committed version of a similar move (Martínez 2013). He argues that homeostatic property clusters are privileged candidates to figure in representational content.[11] Artiga generalizes that view: content is given by a subset of properties which explain why candidate properties tend to co-occur, even when there is no homeostatic property cluster (Artiga in submission).

The problem with all three of these proposals is that they focus on the way the information that the system is responding to is generated: the homeostatic property cluster (if there is one) that underlies the incoming information, or the source of statistical dependencies amongst sources of information. This property need not be the same as the property or properties that constitute and explain successful behaviour. Speaking loosely, an organism does not care what the most informative property is; it cares about what needs to be in place for its behaviour to be successful. For example, consider a rainforest frog that spawns in small pools of water which, in its habitat, are almost all

[11] Martínez (2015) develops and generalizes the view in information-theoretic terms.

found in *Nepenthes* pitcher plants. The frog recognizes spawning locations by detecting the sight, smell, and typical locations of *Nepenthes* plants. The property underlying this regular statistical structure is the presence of the genus *Nepenthes*. However, the success of its spawning behaviour just turns on finding a suitable pool of water. Spawning in a pool that happened not to be in a pitcher plant would not count as a failure. My theory implies that the frog is representing the location of water rather than the location of *Nepenthes* plants.

In short, while taking much inspiration from earlier teleosemantic treatments, my account departs from them in important ways.

6.3 Compositionality and Non-Conceptual Representation

An important feature of the representations found in the human belief-desire system is that they make use of concepts. Concepts are reusable elements which do not make claims or set goals taken individually: they are unsaturated. Only when put together do they form a saturated representation with a complete correctness condition or satisfaction condition. This book does not attempt to deal with concepts and how they get their content. Concepts do, however, have several features which are also found in some of our case studies: semantically significant constituent structure, unsaturated components and (limited) generality.

I reserve 'concept' for the unsaturated personal-level representations that are expressed in language and combine to form beliefs and desires.[12] 'Non-conceptual' covers all representations that not are concepts or constructed out of concepts. Therefore, all the representations in our case studies are non-conceptual, although as we will see some share some features of conceptual representations.

Concepts obey a wide-ranging generality constraint: they can be recombined liberally with other concepts in the thinker's repertoire. For example, any one-place predicative concept, F, can be combined with any singular concept, a, to produce a saturated representation, Fa, which is a candidate for belief. If the thinker can also think Gb, then they have four reusable components, so for example they automatically have the capacity to think Fb.

I will use the term 'saturated' so as to include non-conceptual representations with a complete correctness condition or satisfaction condition, whether or not they are constructed out of unsaturated elements. So, for example, an output node in the simple feedforward connectionist network in §4.3 is a saturated non-conceptual representation even though it has no semantically significant constituent structure (its complete correctness condition is: *the object encountered is in category A*).

[12] This is related to the 'state view' rather than the 'content view' of non-conceptual representation (Byrne 2005).

Predication is involved when unsaturated concepts are put together to form a saturated representation. Predication is absent from most of our case studies, with the exception of offline use in the rat hippocampus. However, many of the case studies do exhibit semantically significant constituent structure of a simpler kind. They also exhibit some local recombinability, and thus some limited generality. None meets the kind of wide-ranging generality constraint met by concepts.

Consider the visual system which detects plaid motion (§4.7). One layer represents chromatic properties at locations in the visual field, another layer represents motion direction at locations. This gives the system a limited kind of systematicity: for each location, it can represent that location as having a range of colours and it can represent that location as having a range of motion directions. But this is the systematicity of a list. There are no a singular terms representing locations and nothing acts as a recombinable representational constituent. Nor are the colour and motion representations tied to the same vehicle. If the vehicles representing motion direction in one part of space were selectively lesioned, the system would retain the capacity to represent colours at those locations. Each layer independently forms saturated representations about colour and motion respectively.

Now consider the case in §4.6: a single distributed representation in prefrontal cortex (PFC) represents both the colour and the average direction of motion of an array of moving dots. This vehicle represents colour and motion at the same time. The well-known bee dance example is similar: a single dance represents both the direction and the distance of a source of nectar. In both cases the system exemplifies a limited form of systematicity. A range of direction representations can be combined with a range of distance representations. But they do not involve anything like predication. There are no unsaturated components, that contribute to the semantic value but fail to have a correctness condition on their own. If the dimension that goes with colour is removed in the PFC case (as it is effectively, in direction choice trials), the remaining dimension still represents that the array is moving in a certain direction. If the number of waggles were indistinct or ignored, a bee dance would still represent the direction of a nectar source. Each dimension is acting like an independent saturated representation with a complete correctness condition.

An important question to ask is whether a representation has semantically significant constituent structure. The plaid-motion system has two different representational vehicles, neither of which has semantically significant constituent structure. The PFC colour-motion system has a single vehicle with two semantically significant dimensions of variation. That is semantically significant constituent structure. There are a range of syntactic states each of which can represent both the colour and the average direction of motion of a stimulus. The correctness condition for these representations is something like: *the currently presented array is of colour abc and is moving in direction r*. Two elements of that correctness condition correspond to two dimensions of variation of the vehicle (colour and motion direction). Other elements of the correctness condition do not correspond to any dimension of variation in the representational vehicle: e.g. which stimulus bears these properties and when.

The ability to vary two features independently is an important kind of semantically significant structure and a notable source of the representational power of that system in the PFC. But it is important to distinguish this from having unsaturated constituents. In the PFC case and the bee dance case, neither dimension of variation is predicated of the other. Each on its own is capable of making a saturated claim. The human conceptual system by contrast makes use of unsaturated elements and predication.

Unsaturated elements can arise when there are multiple dimensions of variation corresponding to different features and there is no stabilization story to be told about how each could prompt behaviour independently. Conditions for successful behaviour only arise when all are tokened together. (Perhaps no behaviour is produced at all when one dimension of the vehicle is tokened on its own, or such behaviour as is produced played no role in stabilizing the mechanism.) Offline place cell activation may be like this (§5.7b). Nothing functional follows from the activation of a single place cell offline in isolation. Co-activation of two or more place cells is required for the offline system to contribute towards the system's task functions. In such cases neither vehicle, tokened on its own, has a complete correctness condition. It is only when two place cells are active that the relation of co-activation is instantiated. The activation of the two place cells then forms a representation with a complete correctness condition (e.g. *location 1 is near location 2*). One explanation for this is that offline place cell activation is unsaturated: each cell contributes a location, and only co-activation has a complete content. (I want to remain cautious about whether this is predication in the sense in which sentences involve predication, or whether it is a different way in which components can be unsaturated—a different kind of function application.[13]) A second explanation is that offline activation of a place cell has suppositional content. It says something like *suppose you were at location 1*. I explore that idea further in the next chapter, when we look at descriptive, directive, and other modes of representing (§7.5b).

The third notable feature of the human conceptual system is its compositionality: any of the unsaturated representations can be combined with any other of the right kind to form a saturated representation. That is, the compositionality of representational vehicles leads the system to obey a wide-ranging generality constraint (Evans 1982), i.e. to exhibit systematicity (Fodor 1987b). The honeybee nectar dance has two semantically significant dimensions (corresponding to distance and direction). Any value of one can be combined with any value of the other. This is only a very limited kind of systematicity. The PFC colour-motion case also displays this kind of limited, domain-specific systematicity. The distributed PFC representation can combine any

[13] One obvious difference is that the predicative element (the relation of co-activation) cannot be tokened without tokening the singular terms. A natural language predicate (e.g. 'red') can be tokened without tokening a singular term.

claim about colour with any claim about motion. Furthermore, given its flexibility, the PFC is likely to have the capacity to also to represent other objects, properties, features, and events. But this is not the wide-ranging systematicity of concepts, where any representation can be put together with any other. The cognitive map and even the PFC system meet only a limited, domain-specific generality constraint. But this is a step in the direction of the full-blown generality constraint obeyed by concepts.

Millikan claims that time and place of production are constituents of simple signs like the honeybee nectar dance. Time and place may figure in the correctness condition, but as with some other elements in the correctness or satisfaction condition of a non-conceptual representation, there is no syntactic type in my sense which corresponds to time or place. For example, there is no singular term picking out location in the way activation of a place cell picks out a location in the rat's cognitive map. Variations in a vehicle are semantically significant when they can take a range of values and their variation makes a difference to downstream processing and/or behaviour. The representational theory of mind is based on the idea that aspects of a vehicle are being exploited for the relation they stand in to features of the environment. Where the mechanism is not able to do different things for different times of tokening, but operates in just the same way whenever the vehicle is tokened (as in the PFC and bee dance cases), time of tokening is not a semantically significant aspect of the representation. Indeed, it is hard to see how time of production could be causally effective in downstream processing unless it is marked or measured in some way.

Another important feature of these cases also sometimes gets the label 'systematic', which is that they form an organized sign system (§5.5; Godfrey-Smith 2017, p. 279). There is a straightforward mathematical relationship between a dimension of variation of the vehicle and the content represented. More activation along one dimension represents a greater amount of motion. Similarly, there is a straightforward mapping from direction of the bee dance to the direction of nectar. Learning or evolution produced a mechanism that follows the mapping. As a result, intermediate values, which may never have been exemplified during learning or evolution, come to have appropriate contents. That, too, could fairly be called a kind of systematicity: there are facts about the mechanism as a whole from which it follows that novel representational vehicles carry determinate contents. Having a single mechanism that can respond to a range of cases in a systematic way is also doubtless an advantage for the system. However, the phenomenon of intermediate values having contents is importantly different from the power that comes from the ability to recombine different representational components, particularly the power of being able to do so in a very general way, as with human concepts.

Representational content derives in part from the situations in which a representation is formed and then acted on to produce behaviour. We saw that the way analogue magnitude representations are produced and used by multiple systems gives them considerable determinacy and establishes reference to numerosity rather than other closely related properties. When we get to concepts, we have syntactic items that are reused across a wide range of situations as they are combined with other concepts. This gives a wide range of uses involved in fixing their contents—giving them scope to have more specific contents. For example, the concepts HOPE and WANT are used across a wide range of circumstances, for understanding others' behaviour and planning one's own. That is part of what allows them to refer to different but closely related psychological properties. The representations in our case studies are not deployed across such a wide range of uses and so are likely to have less determinate contents than conceptual representations have.

To recap, I have picked out three features of conceptual representations and shown how they each occur in some way in some of the non-conceptual representations in our case studies: semantically significant constituent structure, unsaturated constituents, and (limited) generality. The simple feedforward connectionist system (§4.3) and our simplified version of the visual mechanism for detecting plaid motion (§4.7) involve only representations without semantically significant structure. In both cases the system can token more than one representation at once, but these are separate vehicles—what I have called the systematicity of the list. The PFC colour-motion system (§4.6b) and the honeybee nectar dance exhibit semantically significant constituent structure. A single representation has two independent dimensions of variation, each a saturated representation with a complete content. They do not have unsaturated constituents. When place cells are used offline to calculate shortest routes they arguably function as unsaturated constituents, combining so that their co-activation represents spatial proximity. Finally, the systematicity involved in the place cell, PFC colour-motion and honeybee nectar dance representations means that each exhibits some limited domain-specific generality. None of these systems obeys the kind of wide-ranging generality constraint met by concepts.

6.4 Objection to Relying on (Historical) Functions

(a) Swampman

Perhaps the most prominent objection to teleosemantic theories of content targets their defining characteristic: relying in part on etiological functions to fix content. Etiological functions depend on history: a history of selection, learning, or other interaction with the environment. My accounts of content face this challenge because task functions depend partly on history, and task functions play a role in fixing content.

The challenge is made vivid in the literature by considering a 'swampman'—an intrinsic duplicate of a human, but one who arises by complete chance as a result of lightning striking a swamp. Swampman would look and behave like a person with mental states, but any theory of content relying on a historical notion of function will imply that he does not have mental representations, states with content, at least at the moment of creation. Where task functions are based on natural selection, only a system that is the result of selection will thereby have content; where task functions are based on learning or contribution to persistence, a swamp system will not have content until it has undergone some interaction with its environment involving learning or helping the organism to persist. Chapter 3 set out that consequence, illustrated with a toy example (§3.6). This section (i) offers a positive argument that this is the right approach, and (ii) compares my response to others in the literature. I will leave aside task functions based on deliberate design, but they too require history: that a system has been designed or co-opted for certain functions.

We could imagine a swamp system that is an intrinsic duplicate of any of the cases set out in Chapters 4 or 5. The swamp system would have the same behavioural dispositions, so would have robust outcome functions. For example, a swamp duplicate of the system in §4.7 would have a disposition robustly to catch a partly obscured moving object producing plaid motion. It would do so making use of a structure of internal processing, where those internal elements stand in appropriate exploitable relations to distal features of the environment. Since there are robust outcomes involving distal objects and properties, which proceed via a multitude of different proximal routes, there will be distal-involving real patterns in the way the object would interact with its environment, patterns that do not depend on history (which is what I relied on in §3.6). If content did not depend on stabilized functions, but only on the robust outcome function aspect of task functions, then content would inhere in the swamp system. Why isn't that a perfectly good notion of content?

Recall the distinctive 'explanatory grammar' of representational explanation: correct representation explains successful behaviour and misrepresentation explains failure. It is because the success or failure of actions does not depend just on intrinsic properties of the organism or its bodily movements that I argued that this explanandum called for explanation by relational properties of the system (here, relational properties of internal components of the system). What I want to argue now is that, without appeal to history, in the simple cases we are considering here, there are no other ingredients to draw on to make it the case that some consequences should count as successes and others not. That is, the thing which contents are called on to explain—success and failure of behaviour—is absent in a simple system that lacks history (cp. the 'no explanandum' argument, §1.5).

Consider a swamp system corresponding to the plaid motion object catcher in §4.7—call it 'Catcher'. Compare Catcher to another swamp system that happens to have the robust disposition to reach out just to the outside of the direction of motion of a moving object, so that the object bounces off the edge of its hand and passes by; call it 'Misser'. Misser robustly achieves the outcome of glancing the edge of its hand off passing objects, and will do that from many starting positions, adjusting in real time for perturbations in the path of the object. If content were founded just on robust outcome function plus appropriate internal mechanism, both Catcher and Misser would have content. Catching the object would count as a success for Catcher, and were it occasionally to miss, that would count as a failure. The converse is true for Misser—the occasional catch would count as a failure.

An appropriate tweak of internal workings would turn Catcher into Misser. Suppose that too happens by chance. If Tweaked-Catcher now interacts with an object and drops it, is it successfully achieving the same robust outcome function as Misser, or is it misrepresenting the trajectory of the object and so failing to achieve the robust outcome function it had before it suffered the tweak? If content were founded just on robust outcome functions, then whatever the system is disposed to achieve robustly would count as success. So, Tweaked-Catcher would not be misrepresenting, but would be successfully performing the same robust outcome function as Misser. A swamp duplicate of a human that happened to be disposed robustly to pick and eat a type of berry which is poisonous would count as behaving successfully, even if it would soon die or learn to avoid the fruit. We want our theory to allow that there are cases where error leads a system to pursue a poor outcome robustly; for example, a guided missile that systematically misrepresents its location and so robustly arrives a kilometre north of its target. There is no room for such cases if content is based just on current robust outcome functions. That notion of function does not furnish the resources to constitute some robustly produced outcomes as genuine successes and others as failures.

An approach that builds history into the notion of function can make this distinction. Task functions are established by the convergence of stabilized function with robust outcome function at the time of stabilization. If damage or a tweak to the system alters its robust outcome dispositions, then it will be robustly disposed to produce unsuccessful outcomes—outcomes which are not amongst the task functions of the system. Indeed, it is a strain to apply the notion of success just on the basis of robust outcome function. For a moderately complex system like the one in §4.7, very many different outcomes could be robustly produced through small changes to the internal operation of the system. With such a wide array of outcomes potentially counting, it seems tendentious to label each a potential way of distinguishing successful from unsuccessful behaviour. Too many behaviours potentially count as successful. What is missing is any connection with doing good for the system—the sense that successful behaviours are ones that are or have been beneficial. It is the historically based notion of stabilized function that makes success, as constituted by task functions, retain a connection with goodness or benefit.

That is just an intuition, but it mirrors the argument in Chapter 3 which was based on the underlying motivation for representationalism. Representations get their explanatory bite in these simple systems because there is a real cluster in nature where selection, learning, and contribution to an organism's persistence go along with having dispositions to produce certain outcomes robustly, and with doing so by having internal processing that exploits relational properties of internal components. Severing the connection between function and some kind of consequence etiology takes us outside that cluster. Similarly, a forward-looking notion of benefit or consequence is not one of the items that found the cluster. (Recall also the positive arguments against forward-looking accounts offered in §3.4d and §3.7.) It is the existence of a consequence etiology in the past, even the very recent past, which goes along with producing certain outcomes robustly—not the fact that those outcomes would (or might) lead to good consequences for the system in the future.

That is an argument for keeping consequence etiology, hence history, in the picture. The kind of history that counts may be very recent. In most of our case studies the stabilized function is based on a history of learning, and is not derivative from evolutionary history. As soon as a swamp system starts interacting with its environment and learning, it will rapidly acquire task functions. So, it won't be long before there is a basis for counting some outcomes as successful and others as unsuccessful, and then we can start explaining the success and failure of its behaviour in terms of correct and incorrect representation.

Similarly, a swamp human will start with only as-if memories but will soon acquire genuine memories of its interaction with the swamp. It will start by having only an empty simulacrum of relations with other people, but will soon start building up friendships with the people it interacts with. The swamp human is importantly disanalogous to swamp versions of our simple systems, since the extra sophistication of its cognitive apparatus, and/or the fact that it is conscious, may make it a genuine representer from the moment of creation. But the analogy serves to illustrate that it is not unusual that mental properties should depend on interaction with the environment and build up very quickly in a swamp system. When we turn to our case studies, like the reaching system in §4.7, a small amount of interaction with falling objects, with feedback serving to fine-tune the system's dispositions, would be enough to constitute catching objects as a task function of the system. So, it wouldn't be long before a tweak to the system which turned a Catcher into a Misser would count as a failing, with unsuccessful behaviour in the latter rightly blamed on its disposition systematically to misrepresent the location or trajectory of objects in its environment.

In short, my answer to the swampman challenge—the objection to content being partly historically determined—is to cut down the scope of the objection and then accept the consequence, offering a positive argument that, in these simple systems, content should depend partly on history. The scope of the objection is curtailed in two ways. First, because my view does not imply that a swamp human would lack contentful conscious states or thoughts. Content of personal-level representational states may indeed be fixed ahistorically. Secondly, because even in our simple case studies, some contents would soon be established once the system has had a chance to interact with its environment.

(b) Comparison to Millikan and Papineau

My answer to the swampman challenge is different from those previously given by Millikan and Papineau, but not radically so. Millikan argues that a swamp creature would not be part of a real kind, and so generalizations from real humans to swamp creatures would be unsupported (Millikan 1984, pp. 93–4; 1996). They could at best be right by accident. Humans are part of a real kind, but that is a historical kind, the species *Homo sapiens*. According to popular cladistic views of biological classification, species are constituted by shared descent, not by any current property of a population of organisms like shared DNA.

The trouble with this response is that it does not tell us why generalizations about content should go with the historical kind *Homo sapiens*. Human organisms are also physical objects, and as such they obey the laws of gravity. Those generalizations apply to them on the basis of membership of a currently constituted category. So generalizations about how a human falls from a cliff would unproblematically extend to swampman. Why should content properties go in the historical camp? That is particularly puzzling given that swampman looks and talks as if he is susceptible to non-historical generalizations based on attributing representational states in standard ways. And it is agreed on all sides that expectations about behaviour, so formed, would be fulfilled with swampman just as they would be for his real human doppelganger.

Varitel semantics gives us the resources to explain why the generalizations that found content do not extend to swamp duplicates of the systems in our case studies. Duplicates do not fall into the pattern whereby robustness of outcome goes together with internal workings, exploitable relations, and a history of selection, learning, or contribution to an organism's persistence. We could do as-if representational explanation, to some extent, with systems that do not fall into this cluster. But when that worked, it would not be because they exemplify the collection of properties that give content-based explanation its bite. We can still, of course, explain their behaviour in terms of internal workings and how those components are affected by inputs so as to give rise to outputs. As with explaining the trajectory of a falling human in terms of physics and gravitation, that would be to move to a different kind of explanation. It is not because *Homo sapiens* is a historical category that content-based generalizations are inapplicable to the swamp system. It is because the cluster that makes content-related properties project to new cases better than chance is not present. If we do project from real systems to swamp systems, we are relying on currently constituted properties that are much more widely applicable—that apply very liberally in the natural world—and our explanation correspondingly has considerably less explanatory purchase.

Papineau answers the swampman case in a different way (Papineau 2016). He argues that the etiologically based account of content is an a posteriori reduction of our everyday notion. It captures the features which, as a result of scientific discoveries and philosophical theorizing, we have discovered to be important for tying our everyday notion together. If there were lots of swamp creatures around, then our explanatory practices would turn on something else. But in the actual world our explanatory practices are based on the existence of historically constituted properties. Papineau concedes that we could explain and generalize in terms of current properties, and that such generalizations would apply to swamp systems, but he argues that nothing would be gained in the actual world thereby, since swamp systems do not occur here.

My approach is in the spirit of Papineau's observation about an a posteriori reduction. But he shouldn't concede that equally good generalizations in terms of current properties are available. If there were simply a tie between the explanatory power of currently constituted compared with historically constituted contents, then the absence of swamp creatures in the actual world would not be a decisive consideration. Either pattern would be available as the basis for prediction and explanation. So, I think Papineau's argument needs to be supplemented with the observation that the currently constituted properties that are available for explaining the behaviour of actual creatures and swamp creatures in a unified way are much less satisfactory. They hook on to patterns that exist in those creatures, but that are also found much more widely in nature, and in various degrees. The distinctive explanatory purchase of representational explanation arises because there is a more tightly delineated cluster of properties that arises when consequence etiology gives rise to robust outcome functions supported by internal workings. Swamp creatures fall outside that pattern—if we are restricted to current properties, we cannot explain their behaviour, in this characteristic way, by reference to it.

6.5 Norms of Representation and of Function

(a) Systematic misrepresentation

Another major line of objection to teleosemantic theories is that they do not deliver the kind of normativity that is characteristic of mental content. As I have characterized representational content in the chapters above, the difference between correct and incorrect representations can be captured descriptively. It is a descriptive difference to which norms can readily be applied, just as whether a friend waves or not when we see them is a non-normative fact which can be the basis of praise or censure. Normative properties are not an inherent feature of content. Misrepresentation is one way of explaining a failure to perform task functions. So, if we took it to be a good thing that an organism should fulfil its biological functions, then there would be something wrong with misrepresenting when doing so produces behaviour that fails to fulfil its task functions. But biological well-functioning is just another descriptive distinction. It does not bring genuine normativity into the picture. A descriptive distinction to which norms can be applied is all we should expect in the kinds of cases we have been considering. Norms in a stronger sense—connected to what one ought to do—may arise for representations that are connected to language use or otherwise embedded in a social context, but that is not in play here.

Critics have argued that teleosemantics mistakenly elides misrepresentation with malfunction. That objection needs to be taken seriously even if both distinctions are in the end purely descriptive. Since it is sometimes in an organism's best interests systematically to misrepresent how things are in the world, the objection runs, correctness of a representation cannot be equated with promoting fitness, or indeed with any kind of biological well-functioning. My answer comes in two parts. As we will see shortly, varitel semantics does not equate misrepresenting with malfunctioning. It allows for malfunctions that are not caused by misrepresentation, for misrepresentations that do not lead to malfunction, and also for misrepresentations that are produced systematically in the organism's evolutionary interests. However, Peacocke puts forward a case which suggests that there is a deeper gulf between misrepresentation and malfunction than my account allows. My answer to that is that we have no reason to think that such cases arise in the kind of subpersonal systems dealt with by varitel semantics. I deal with that first.

Peacocke's example is a case in which a creature systematically misrepresents a predator which is 30 feet away to be only 20 feet away (Peacocke 1993, pp. 224–5). The creature runs away faster as a result and gains a selective advantage by doing so. In that particular example, if there were no other behaviours involved in fixing content, and if the flight response at that speed had indeed been the best trade off of costs and benefits for predators at 30 feet, then Millikan's theory implies that the content is that the predator is 30 feet away (i.e. that it is not misrepresenting). If the details were filled in a bit more so that my account in Chapter 4 applied to this case, then it would have the same result.

These examples do however typically assume that the representation in question is involved in some further pattern of behaviour which fixes its content (see Figure 6.1).[14] That could certainly be the case when we get to human beliefs and desires, which may offer an explanation as to why, in their verbally reported explicit beliefs, human subjects systematically over-estimate the efficacy of their own actions (in contrast with so-called 'depressive realism': the more accurate estimates typically offered by people with clinical depression: Moore and Fresco 2012). If behavioural dispositions to act on a set of representations are formed in one context, and are relatively developmentally fixed, then it may make sense to 'trick' the system when deploying it in other contexts, if the behaviours appropriate to the new context would be different.

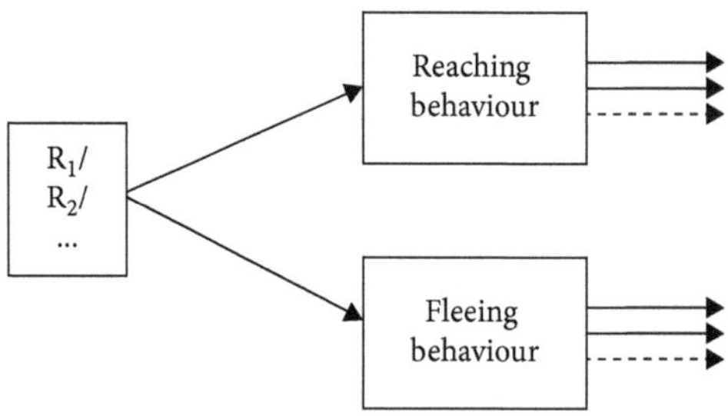

Figure 6.1 The structure of the case from Peacocke (1993, pp. 224–5).

[14] Peacocke mentions another behaviour prompted by these representations: throwing a stone aimed at 20 feet. That would indeed fix a different content, although it is hard to see how the two behavioural dispositions would evolve at the same time. It is more likely that one or other behaviour falls outside the pattern in virtue of which selection has occurred, in which case there is likely to be a directive content which goes unsatisfied, given that the descriptive representation of the situation correctly represents the target predator as being 30 feet away.

However, our simple case studies do not have that kind of structure. In our cases a system has been stabilized for doing a range of things in one context. There is not a second context where benefit is pulling in a different direction so that two different kinds of correctness can become established. Where there are two different routes to behaviour, each of which has been stabilized in different circumstances, then two contents can arise, and they can conflict.[15] Where there is just one route to behaviour, the kind of conflict between correctly representing and well-functioning pointed to by Peacocke does not arise. At least it has not been shown that the intuitive case for the challenge, based as it is on thought experiments where there is more than one route to content, can be extended to our simple cases.

A reason to think that it cannot is that, without further articulation, content ends up being fixed so as to align with whatever story is told about selective or evolutionary benefit. Here is an analogy. Representation theorems in decision theory fix a person's degrees of belief based on the pattern of their choice behaviour. This makes it impossible for a person whose choices obey some basic principles to systematically misrepresent subjective probabilities. A different pattern of choice behaviour would imply a different pattern of degrees of belief. It is only if there is further structure, for example a system of explicit, conscious beliefs expressed verbally by the subject, that those beliefs can come to misrepresent systematically, which in this case would be to say that the explicit beliefs differ from the probabilistic beliefs attributable to the person on the basis of their non-verbal behaviour.

Although I reject the kind of radical disconnect between correctly representing and biological functioning suggested by Peacocke's case, varitel semantics does not elide misrepresentation with malfunction or equate the two. Representational content calls for a constellation of factors, amongst which task functions are just one component. Representing correctly does not reduce to performing task functions successfully, nor does misrepresenting reduce to malfunctioning. (This is the source of the disagreement with Neander discussed in §6.2h above.) A behaviour may have unsuccessful consequences even if the organism is representing everything correctly, for example if something goes wrong with the execution or the environment is uncooperative. Conversely, an organism can misrepresent but through luck produce entirely successful behaviour in accordance with its task functions. Neither distinction reduces to the other. Furthermore, varitel semantics allows for systematic misrepresentation. The costs and benefits of behaviour may be such that selection or learning is tuned to produce many false positives: to be liberal in representing p when p might be the case (Godfrey-Smith 1989, 1991). Then the organism will often be misrepresenting, and acting unnecessarily, although it is operating in just the way evolution has designed it to.

[15] Corollary discharge is in one sense like that (§4.5, §7.4), although there the contents that exist relative to the two different uses are closely related.

In short, varitel semantics allows for some kinds of disconnection between correct representation and biological well-functioning, and there is no reason to think that a more radical disconnection could arise in the kinds of subpersonal systems it targets.

(b) Psychologically proprietary representation

The role of normativity also marks a deep difference between my account and the theory put forward by Tyler Burge in *Origins of Objectivity* (Burge 2010). Burge presents an account of the nature of perceptual representation which, although based firmly in the natural sciences, is very different to the approach I have adopted here. Burge presents his theory in opposition to naturalistic approaches that reduce representational content to a combination of information and function. His first argument against these approaches is that misrepresenting is not necessarily counter to an organism's fitness or biological interests. We have just seen how I address that concern.

Burge's second argument is that teleosemantic accounts are too liberal, setting the border of intentionality too low, and thus allowing in cases where content has no real explanatory value. I disagree. My account of the explanatory purchase of representational content is based on an externalist explanans (one which carries a distinction between behavioural success and failure), together with an explanandum that appeals to externalist properties of internal vehicles (§2.3). It calls for instances of the natural cluster identified in Chapter 3 (§3.2), which is not unduly liberal. But it is true that this account does not depend on the representing system being particularly sophisticated, so it does potentially apply quite widely. This does not set the lower border on representation too low, however. Content-based explanation does indeed have explanatory purchase in this wide class of cases, as I will argue in Chapter 8 (§8.2, §8.5).

Burge makes a third, related argument: that representation proper is proprietary to psychological systems, and is normative. Our accounts of content are thus inadequate for two reasons. They do not capture something distinctively psychological—varitel semantics can in principle apply to non-psychological systems. And they account for content in non-semantic, non-mental, non-normative terms, thereby missing the constitutively normative nature of mental representation.

Given my pluralism, I'm happy to allow that the content of some kinds of mental representation might be importantly different from the subpersonal and from non-psychological cases. Burge may well be right that something different and more sophisticated is needed to characterize the content of personal-level perceptual states; and indeed of beliefs and desires. And there may indeed be something normative going on there, given the way thought and language is embedded in social practices. There is clearly an important difference between our views, however, since I argue that my approach is adequate to account for some genuinely psychological representations, the subpersonal representations in my case studies. I see no good reason to think that subpersonal representational content, of the kind widely relied on in experimental psychology, cognitive neuroscience and the other cognitive sciences, is a kind of content that is proprietary to the psychological. Indeed, we have good reason to think that it exists more widely, in computational systems that have the same functional profile. Therefore, given our explanatory target, I reject the need for a psychologically proprietary account.

What of normativity? Burge's approach is deliberately non-reductive: he characterizes what it is to be a representation in terms of having correctness or veridicality conditions, which he characterizes in turn in terms of being a representation, rather than a mere sensory state or informational registration. It is important for Burge that perceptual representations show constancy effects (they are formed by a many-one mapping from sensory inputs). But he doesn't in the end want to characterize that many-one mapping causally, but to do so in normative terms. A state that is formed in a common way in response to a variety of inputs would not count as showing a constancy unless it had genuine veridicality conditions (i.e. unless it were a representation). Burge is happy for his account to contain this tight explanatory circle because he rejects the need to 'naturalize' representational content. He argues that the notion of representation does not need naturalizing, if that requires that representation be explained in other terms. It is an entirely un-mysterious property that plays a central role in the successful science of perceptual psychology and is fully vindicated thereby.

Burge's thought here would be, I think, a good response to the claim that we should doubt whether there are any representations. Their central role in the sciences of the mind gives us good *prima facie* evidence that there are representations. But my project is directed at a rather different problem: not of showing that there are representations, but of trying to understand their nature better. Burge's theory does something to characterize their nature, by locating them in a small local holism involving correctness, constancy, and content, but my hope is that we can do more. An account of representational content in terms that are non-mental, non-semantic, and non-normative tells us considerably more about their nature. Of course, it could have turned out that Burge's account is all that can be said to illuminate the nature of representation. But for the kinds of cases discussed in the foregoing chapters, that would be to give up too soon, since more illuminating accounts of content are available.

6.6 Conclusion

Varitel semantics has several resources for dealing with indeterminacy. It gives rise to more determinate contents than informational semantics or consumer-based teleosemantics. The remaining indeterminacy is a virtue: it is what we should expect in simpler systems with fewer interacting components. Some of the non-conceptual representations in our case studies exhibit some features exemplified by concepts: semantically significant constituent structure; unsaturated components; and limited, domain-specific generality. However, they lack the wide-ranging generality of personal-level concepts. Since the content of a concept is fixed by reference to a wide range of uses, in combination with many other concepts across many different contexts, conceptual content is likely to be more focused, and thus more determinate, than the contents determined by the simpler interactions and recombinations exemplified in our case studies.

Task functions import a historical component into content determination. That is needed to bring the explanandum into view, the explanandum to which representational explanation is directed, namely explaining successful and unsuccessful behaviour. A history of stabilization operates to constitute some outcomes as being successes—beneficial results—and others as failures, so that even outcomes that are now robustly produced can count as failures in some circumstances. So, we should not expect representational explanation to get a grip, in these simple cases, unless the system has some history of interaction with its environment. However, after even a short period of interaction, some task functions, hence some contents, will begin to be established. So, historically based functions play an ineliminable role in the accounts of content advanced in previous chapters. Varitel semantics does not reduce misrepresentation to malfunction. Misrepresentation does not imply failure to perform a task function, nor the converse.

In short, varitel semantics does a reasonable job of addressing the standard challenges in the literature.

7

Descriptive and Directive Representation

7.1 Introduction

The representations we have discussed so far have in fact come in two distinct kinds: those with descriptive content and those with directive content. Speaking informally, descriptive representations are supposed to match the way things are in the world and are correct or true when they do. Directive representations are also associated with a worldly condition—a condition they are supposed to bring about. A directive representation is satisfied when its condition comes to obtain. This chapter engages with the long-standing philosophical debate about how that distinction should be drawn.

It has proven difficult to characterize the distinction precisely (Humberstone 1992) and I don't aim to do so in a theory-neutral way here. However, there is a clear distinction, in the right neighbourhood, which can be made within the varitel framework. Section 7.2, following some preliminary points of clarification (this section), gives my account of the distinction. Section 7.3 shows how the distinction works in our case studies. Section 7.4 compares my way of drawing the distinction to others in the literature. Section 7.5 briefly considers some further kinds of sophistication that arise in the case studies.

Many previous treatments have used the terms 'indicative' and 'imperative' for descriptive and directive content. However, those terms are used in linguistics and philosophy of language to label the grammatical mood of a sentence, which need not align with its content (e.g. 'the door is open' is an indicative sentence that can have the directive content *shut the door*). Using 'descriptive' and 'directive' is designed to avoid confusion.

Beliefs and desires, respectively, are the paradigmatic examples of descriptive and directive representations. There are other types of propositional attitude at the personal level; for example, supposing, entertaining, and imagining. Although our case studies are simpler, it is not obvious that they are restricted to only having the descriptive or directive mode of representing. In §7.5b we will see that something like suppositional content may arise in one of our case studies, when representations are used in conditional reasoning.

I have characterized the descriptive–directive contrast as a difference in content. Others would say this is a difference in attitude between states that could have the same content. According to that usage, content is a worldly condition and the descriptive or directive is the attitude in which that content is represented. That picture derives from the model of beliefs and desires, where the same representational vehicle can be deployed in the attitude of belief (put in the 'belief box'), in the attitude of desire (put in the 'desire box'), in the attitude of intention, and so on. An account of the difference between belief and desire is given in terms of a representation with the same content figuring in two different functional roles (Fodor 1987a). These states share content in a restricted sense (e.g. that p), while differing in attitude.

I prefer to use 'content' for the full representational import of a state, so as to include a specification of its mode of representing. Used this way, a correctness condition like *the door is open* and a satisfaction condition like *open the door* are different contents. The functional role of a putative representation fixes this full specification, and it is part of the problem of content to say how that can be so. A theory of content that only delivered content in the more restricted sense would be incomplete. Furthermore, my terminology stops us simply assuming reusability. It's not always the case that the same attitude-neutral vehicle can be reused in different functional roles so as to give it different modes of representing. That is a plausible hypothesis about beliefs and desires, making the content–attitude distinction useful there. However, it is not a general feature of the systems we have been considering. It does arise in certain cases (§§7.4 and 7.5 below). That can be captured by giving a full specification of the content, so as to include the mode of representing, while making clear that the same vehicles are reused in different functional roles, so that representations in different modes deploying the same vehicle involve the same worldly condition.

Another terminological choice is to talk about directive contents concerning a condition C. That encompasses outputs that are movements of the system and actions performed by the system. It is slightly awkward to call the movement of a body part a 'condition' produced by the organism, so it is worth emphasizing that I do intend the term to cover these cases. Limb movements can be individuated in terms of intrinsic properties of the system, but the limb moving that way counts as one type of condition C which the system brings about.

7.2 An Account of the Distinction

The accounts of content offered in Chapters 4 and 5 do not make a distinction between descriptive and directive contents. They home in on conditions whose obtaining or otherwise is important for explaining successful behaviour, without distinguishing between conditions which the system causes to obtain and conditions that obtain antecedently. Recall the central idea: an exploitable relation figures in an unmediated explanation of how a system performs task functions. In some cases that exploitable relation is a matter of receptivity: the system makes use of inputs in order to go into a state or states that stand in the exploitable relation, and the relation is exploited by relying on those states in further processing or by conditioning behaviour on them. In other cases the exploitable relation is a relation to outputs that the system produces: the role of the vehicle in achieving a task function is to bring about a certain result.

Exploitable relations that are part of an unmediated explanation of how a system achieves its task functions can be of either kind. Although the accounts of content do not differentiate between descriptive and directive modes of representing—nor need they do so in order for the obtaining or otherwise of the content-condition to explain success and failure of behaviour—we can supplement them so as to classify whether the exploitable relation plays a descriptive or directive role (or both, or neither).

A tempting first thought is that representations whose tokening is caused by inputs to the system are descriptive and those that cause outputs are directive. And that asymmetry does indeed exist in many cases. For example, the analogue magnitude system carries UE information about the numerosity of collections of objects, that correlation exists because of how the system is sensitive to objects in the world (§4.6a), and the representations do indeed have descriptive content (about the number of objects presented). Motor programs, by contrast, carry UE information about bodily movements and actions caused by the agent (§4.5), and they have directive content.

However, in varitel semantics content is based on explanatory considerations, not simply facts about causal sensitivity and causal effects. So, the descriptive–directive distinction should turn on whether producing condition C or reflecting condition C figures in the explanation (in an unmediated explanation of how—through the representations implementing an internal algorithm—S performs task functions F_j). Furthermore, there is no requirement in the theories above that the objects and properties represented should play any causal role at input. (Recall from §6.2c: the frog's retinal ganglion cell firing could carry correlational information about and represent the location of flies even if the cause of cell firing were patches of light that attract, and so precede, the presence of flies.)

Rather than causation by C, what is characteristic of the descriptive role is that the correlation of R with C, and hence C's obtaining better than chance, enters into the explanation of how downstream effects of R have or achieve their task functions. Contrast the directive case, where what is important is producing C. In the directive case, there would be no point tokening R if C obtained already. In the descriptive case, the whole point of tokening R is that C does obtain already, or at least that it is likely to obtain at the point when behavioural outputs prompted by R occur.

A complication arises because there are cases of corollary discharge (§4.5) where a (directive) motor program acquires an additional descriptive content (having both modes of representing, it is a *pushmi-pullyu* representation). As well as having the job of causing a bodily movement, it also has the job of telling other subsystems that the movement is about to take place, so that they can produce appropriate outputs, such as balancing movements. To deal with this complication it is easier to define the directive mode of representing first.

> *Directive Content (based on UE information)*
> For internal component R carrying UE information about condition C in a system S with task function or functions F_j:
> if R's producing C is part of an unmediated explanation of S's performance of task functions F_j,
> then R has the directive content: *produce C*

Recall that we defined the explanandum, 'S's performance of task functions F_j', as including both the question of how outputs F_j have been stabilized (through evolution, learning, or contribution to an organism's persistence) and the question of how they have been robustly produced (§4.2a). Directive contents arise where exploitable correlational information becomes UE information because of R's role in causing a condition C to obtain. R's role in the mechanism that was stabilized and produces outputs robustly was to cause condition C to obtain.

We should recall a subtlety discussed in Chapter 4. Output-based UE information may concern the means by which a task function is brought about, as with a motor program that is used to bring about a task functional output F. But there are also directive representations that concern task functions directly, such as *get sugar*. So then the internal computations involve two directive representations: one that selects a task function to be achieved in the current context (*get sugar*) and another that programs the means to that end (*move the right hand to (x,y,z)*).[1] The correlation in the former case is simply a matter of production of a task functional outcome, rather than an explanation of how the system produces that outcome. We saw in §4.2a that the output correlation can nevertheless form part of an explanation of how the whole system comes to produce this outcome in a way that was stabilized and robust.

Having identified the UE information that is constitutive of directive content, we can now turn to descriptive content. The basic idea is that the explanatory role of a correlation in the descriptive case is to raise the probability that some condition C, relevant to how the system performs task functions, obtains. But an obvious complication

[1] An output that is a means for producing a task functional outcome, e.g. moving the eyes thus-and-so, need not itself be a task functional output, e.g. if it does not meet the robustness condition.

arises. The job of a directive is to bring about a certain output. In doing that it raises the probability that a certain condition will obtain (namely that the output is produced). The directive representation *get sugar* should not automatically be counted as also being a descriptive representation just because its satisfaction condition obtaining (the organism getting sugar) is part of an explanation of how the mechanism was stabilized. So, we need to exclude this kind of case in defining descriptive content. Descriptive content concerns a condition C whose obtaining when R is tokened figures in an unmediated explanation of robustness and stabilization, but not in virtue of R's having the causal role to produce condition C.

Descriptive Content (based on UE information)
For internal component R carrying UE information about condition C in a system S with task function or functions F_j:
if C's obtaining when R is tokened is part of an unmediated explanation, that does not go via R's producing C, of S's performance of task functions F_j,
then R has the descriptive content: *C obtains*

Now we can return to the issue of corollary discharge. While we don't want our definition of descriptive content to automatically constitute every directive representation as also having descriptive content (as a pushmi-pullyu), we don't want to rule out that a directive representation can also acquire descriptive content, in virtue of a second functional role in plays in the system.

Consider a motor command which correlates with the production of a particular bodily movement, holding the right arm out horizontally. That is a role the motor command is supposed to play so it is a representation with directive content. Other motor systems react to this state by producing balancing movements that stabilize the body; for example, tensing muscles in the legs and pelvis (Bouisset and Zattara 1981). These movements are performed before any sensory feedback has come in to report that the right arm has gone up. They are based on the motor program, since it is a reliable sign that the arm will shortly be up, and hence of the need for compensatory adjustments. These compensatory adjustments are motor outputs, and a condition involved in explaining why they are stabilized is that the right arm is up (otherwise they would be a waste of effort and destabilize the body in the other direction). The motor program itself carries the correlational information that this condition is about to obtain. I would argue that *that* correlation figures in an unmediated explanation of how the organism achieves the task function of remaining upright. So, the motor command, in virtue of a second functional role that it plays in the system, also carries the descriptive content that the right arm is raised.

Does our definition of descriptive content allow that there is also descriptive content in this case? The organism performs the task function of staying upright. It does that when the arm is raised by producing balancing movements of the muscles of the legs and torso. It produces those balancing adjustments in appropriate circumstances by relying on an internal vehicle R that correlates with the arm's being raised (condition C). So, R looks to have the descriptive content that the arm will shortly be raised. What about the caveat that the explanation 'does not go via R's producing C'? R does in fact produce C (i.e. cause condition C to obtain). But the way the obtaining of condition C combines with balancing movements to explain performance of the task function of remaining upright does not depend on R's producing C. It would work just as well if the correlation of R with C depended on R's detecting some other sign of C and R had no role in producing C. The correlation does not fall under the exception, so qualifies as descriptive: the motor program does indeed have a second, descriptive content. So, this definition of descriptive content allows that representations can have both descriptive and directive content in appropriate cases, without going too far and entailing that all directives also carry descriptive content.

We can make a parallel descriptive–directive distinction within the account of UE structural correspondence (Chapter 5). In the case of structural representations, a representation is tokened in virtue of two or more vehicles standing in a certain relation within the system. So, a particular representation R_1 is realized when a relation V holds between two vehicles v_1 and v_2 (e.g. one place cell is activated after another). Under the UE structural correspondence which gives the content of the set of representations R_i which includes R_1, the two vehicles v_1 and v_2 stand for two entities in the world, say x_1 and x_2, and the tokened relation V stands for a relation between x_1 and x_2, call it H. R_1 then represents condition $H(x_1,x_2)$.

Directive Content (based on UE structural correspondence)

For representation R_1 from a set of representations R_i bearing the UE structural correspondence to relation H on a set of entities x_k under which R_1 corresponds to $H(x_1,x_2)$,

if the fact that the R_i produce relation H on the entities x_i is part of an unmediated explanation of S's performance of task functions F_j,

then R_1 has the directive content: *produce* $H(x_1,x_2)$

Descriptive Content (based on UE structural correspondence)

For representation R_1 from a set of representations R_i bearing the UE structural correspondence to relation H on a set of entities x_k under which R_1 corresponds to $H(x_1,x_2)$,

if the obtaining of relation H on the entities x_k when a representation R_i is tokened is part of an unmediated explanation, that does not go via R_i producing relation H on the entities x_i, of S's performance of task functions F_j,

then R_1 has the descriptive content: $H(x_1,x_2)$ *obtains*

Consider the spatial navigation system in the rat hippocampus. It exploits the correlation of a place cell with location to tell the rat where it is. Then it makes use of the co-activation structure to run through a series of routes offline until one joins up with a rewarded location, selecting the shortest such route by selecting the shortest or fastest offline sequence. We have been supposing that this structure arises through learning because it has led rats to follow efficient routes to worthwhile locations, food sources say. Getting to previously encountered sources of food are then outcomes F_j that have been stabilized by learning. To explain how they were stabilized we point to the fact that, when the co-activation relation is tokened on place cells, the corresponding relation of spatial contiguity tends to exist on the corresponding locations; for example, that location y is near to location x. The co-activation relation therefore *descriptively* represents that *y is near to x*.

7.3 Application to Case Studies

(a) UE information

In this section I say briefly how my way of drawing the distinction applies to the case studies discussed in previous chapters. It agrees with Millikan's verdict about very simple throughput cases like the honeybee nectar dance. Those intermediates are pushmi-pullyu representations with both descriptive and directive content. Signalling evolved by natural selection and was stabilized because the dances have a tight correlation with the location of nectar at input and with the flight of foraging bees at output.

In the ALCOVE model (§4.3), the output representations are also pushmi-pullyu. The system exploits the fact that they correlate tightly with the category of object presented and with which box they cause the system to place the object in. I argued that the layer of exemplar nodes should not be seen as simply recapitulating with less fidelity the correlations used at the output (since that is less explanatory of how the system works). So, the explanatory correlation at the exemplar layer is with the identity of the exemplar, and not with any particular behaviour. They have descriptive content; similarly for the input nodes. Notice that the pushmi-pullyu representation at output has different (but related) conditions in the descriptive and directive contents (*the object is of type A; place the object in box A*). That is also true of the honeybee nectar dance.

There are also pushmi-pullyu representations where descriptive and directive content concern the same condition. Using motor programs to drive compensatory bodily movements gave us one example (§7.2 above). Corollary discharge will in many cases, for the same reason, have contents like this (§4.5). The predictive comparator mechanisms involved in motor control, discussed in §3.6a, are also somewhat like this in structure (Desmurget and Grafton 2000, Wolpert and Ghahramani 2000), except there the motor content is a pure directive and is transformed before a corresponding descriptive content is represented.

The schematic structure of a predictive comparator mechanism is illustrated in Figure 7.1 A motor command is used simultaneously to drive behaviour and to predict the sensory feedback that is likely to result. The discrepancy between predicted sensory feedback and desired state is used to adjust the motor command even before any

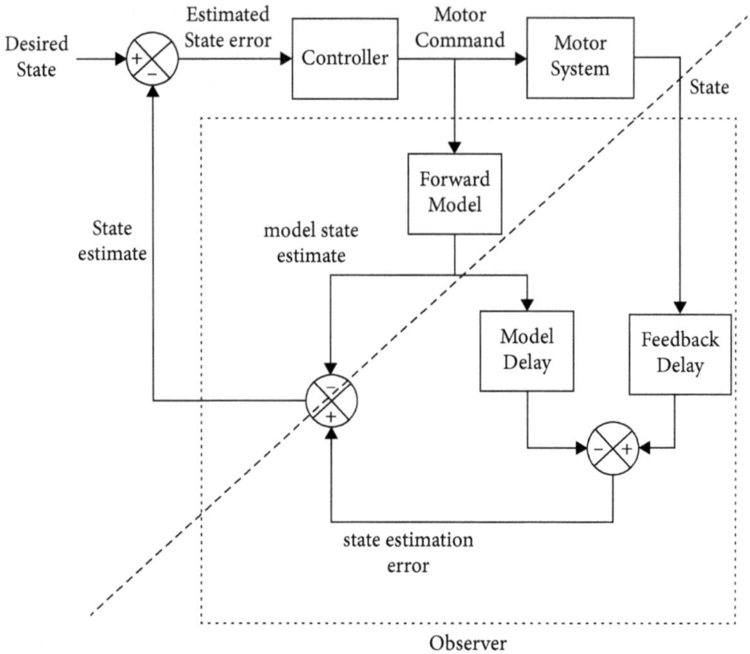

Figure 7.1 A predictive comparator model from Miall and Wolpert (1996, adapted). A copy of the motor command is fed into a mechanism that estimates the likely world state that will result (the 'model state estimate'). That estimate is then compared with the target state ('desired state') so that the motor command can be adjusted even before sensory feedback is received (the processing steps above and to the left of the diagonal dashed line).

sensory feedback has had time to arrive (processing steps above and to the left of the diagonal dashed line in Figure 7.1). However, there the predictive descriptive representation ('model state estimate') is the result of a further processing step after the motor command, so the motor program itself will not have descriptive content. It is a pure directive. The fact that it correlates with outputs means that it can be relied on in order to generate another state with descriptive content about those outputs.

Recall the PFC colour/motion choice system described by Mante et al. (2013) (§4.6b). The computation involves two vehicles at input. One, the context representation, correlates with whether it is currently in a 'colour context' or a 'motion context' (i.e. whether colour or motion is going to be the basis of reward). The other vehicle, call it R_I, has two dimensions of variation, one corresponding to the colour of the stimulus and the other to the motion of the stimulus. R_I is then transformed into a one-dimensional vehicle that drives behaviour (left or right), which is formed by preserving the dimension of R_I that is relevant to choice behaviour on the current trial (colour or motion) and collapsing the other. R_I thus descriptively represents both the preponderant colour and the direction of motion of the current stimulus. The context representations are descriptive representations of whether colour or motion will be the basis of reward in the current context. R_I is transformed into a directive representation R_O which drives an action to the left or right. R_O also correlates with a disjunctive input condition (e.g. *the majority colour is red or the majority direction is left*), but since there are other components of the mechanism correlating more tightly with colour, motion, and choice context, the fact that this internal component correlates with this further disjunctive condition offers no additional explanatory purchase. It does not figure in an unmediated explanation of stabilization. Thus, even in this simple case, there are separate descriptive and directive representations, and no pushmi-pullyu representations.

Pure descriptive contents are found in many of the case studies. Analogue magnitude representations of numerosity are pure descriptives, as discussed in §7.2 above. Representations of individuals in the face recognition system are another clear case (§3.4c). No output behaviour at all need be involved in acquiring a representation of a new individual. The plaid motion detection system (§4.7) descriptively represents motion properties (in thick stripe V2) and chromatic properties (in thin stripe V2). It transforms these into a register R in MT that correlates with the direction of motion of surfaces. The system makes use of that correlation to condition behaviour that is appropriate to the motion direction, so R has descriptive content about the direction of plaid motion. In our stylized treatment we supposed that R is used directly to condition behaviour (a reaching movement in the corresponding direction). If the connection were that direct, R would also have directive content, however in the human visual system behaviour is likely to be highly context-dependent, so no such directive content would arise in MT.

In §4.8 we looked at a simple system that accumulates evidence and computes the probability that various available actions will be rewarded. We explained its operation in terms of exploited input correlations, taking into account probability distributions. Only its outputs have directive content, programming one or other action. This case shows that a vehicle of descriptive content can be activated with various strengths, with stronger activation representing that a worldly condition is more likely. That raises the possibility that directive representations could also come in varying strengths, an idea we will return to in §7.5a below.

(b) UE structural correspondence

I turn finally to our cases of UE structural correspondence. In the rat navigation case, the correspondence between co-activation structure on place cells and spatial structure on the corresponding places is used to calculate shortest routes. The rat doesn't put the locations into spatial relations. The behaviour it performs relies on the correspondence holding when it calculates a route. So, they have descriptive content.

It is harder to find simple cases of UE structural correspondence with directive content. We are familiar with high level cases like using a blueprint to build a building. A blueprint represents the way parts should be put into certain spatial relations. Chris Eliasmith's spiking neuron model for solving the Tower of Hanoi problem uses representations of relations amongst discs and calculates over these relations in working out which disc to move when (Eliasmith 2013, pp. 191–8). These are arguably structural representations of spatial relations. The representation of the target end state uses this structure to represent an arrangement of discs. It is then a structural representation with directive content. There may be lower-level cases involving causal relations. If an organism plans a sequence of actions and represents them with some kind of directed causal graph, which it then consults in order to execute the actions in the right sequence, that causal graph would have directive content about the causal structure of the sequence of actions to be performed.

Dan Ryder's SINBAD model uncovers statistical structure in the world (Ryder 2004, Ryder forthcoming). Ryder claims that SINBAD constructs structural representations. A clever and attractive feature of the model is that it can be inverted so that representations can be used in directive mode so as to drive and guide action. The model discovers statistical regularities in the input, so that it ends up with individual cells that are tuned to underlying causes of statistical regularity (Ryder's 'sources of mutual information'); for example, to natural kinds. When it encounters a series of kinds to which it is tuned, a corresponding series of cells will be activated. Ryder calls the way cell activity mirrors the sources of mutual information encountered a 'dynamic isomorphism'.

The network can also learn which kinds are correlated with which, through forming lateral associative connections. So, if cell B is normally activated by sensory inputs when cell A is, the system will learn that association. Cell A will now activate cell B even in the absence of sensory inputs to cell B, and the converse (in Ryder's model learning produces bi-directional connections). That allows the network to 'fill in' with predictive activity in line with patterns it has encountered in the past. Ryder shows how this network can be taken offline and used in a directive mode of operation (Ryder forthcoming). One cell is specified as a target state. The rest of the network is allowed to adjust, in the absence of input, to reach values that would produce this target state. When the system is operating in this functional role, activity of each node has directive content. Directive content about a target sets off a chain of simple inferences until the system arrives at directive content that it can execute (which thereby acts as an intention or motor program).

As described so far, no structural correspondence between cell activity and conditions in the environment is doing any work. The system is exploiting a collection of vehicles carrying correlational information, both input correlations and output correlations. We saw an example of the way exploitable correlational information about a distal output like getting sugar can depend, for its existence, on an exploitable correlation with a more proximal output; for example, with pressing the left-hand button. Similarly, SINBAD has some exploitable output correlations that depend on others. Once it has undergone learning, the system has a vehicle which correlates with producing a distal outcome, getting hot water, and that correlation proceeds via another vehicle which correlates with producing a more proximal outcome, turning the left-hand tap. Just as input information is transformed through a series of correlational vehicles (e.g. in the perceptual hierarchy) output information can also depend on intermediate transformations. The complete computation is making use of multiple correlational vehicles at output as well as at input, and the way they are computed over instantiates an algorithm for performing task functions. That is not yet a case of structural representation (§5.7a).

A more sophisticated model of the same type might contain structural representations. Suppose the chains of activity between cells correspond, not to co-instantiation of kinds, but to regularly encountered causal sequences. Suppose further, that when the network is used in directive mode, the steps of activity produced by the inverse model are acted on in sequence. T is set as a target, which in directive mode activates A1, then A2, and so on. Rather than connecting all of these to action simultaneously, the system deploys them in serial order, first doing the action driven by A2, then that driven by A1. (To work like this something more sophisticated than the simple bidirectional associations suggested above would have to have been learnt.) Using the network in this way would make use of the correspondence between the temporal order of activation (in the network) and the causal sequence of events (in the world). If this temporal order was important to the reasoning in some way (as described in §5.6b), then it would be a structural representation of the causal order of events.

A notable feature of this case is that there is something like a content–attitude distinction: the same vehicle can be redeployed in descriptive or directive mode. I noted above that this feature, characteristic of beliefs and desires, is not a necessary feature of representational systems in general. It is interesting to see that it can nevertheless arise in a case that is simpler than the belief-desire system.

In fact, the rat navigation system too may take place cells, whose descriptive content we have already captured, and redeploy the same vehicles in a different role which gives them directive content. Consider what happens once the rat has calculated a shortest route and needs to set off. It should follow a path corresponding to the sequence it has just calculated. It does that by putting its place cell system online again, so that its activity reflects the rat's current location. Now let's make a supposition, just to illustrate the point. Suppose that the place cell for the next step in the shortest route sequence becomes active, and the difference between that activity and the currently active place cell is calculated. If the rat moves so as to minimize this error signal, it will tend to move to the next location, where the process is then repeated. When active as a target, the place cell has a directive (correlational) content: *move to location y*. When activated by an input signal, a signal which is changed by movement, a place cell has descriptive (correlational) content: *I am at location x*. The same cell has directive or descriptive content about the same location, depending on its current functional role. (Note that the representational *relation*—co-activation—has descriptive (structural correspondence) content throughout.)

7.4 Comparison to Existing Accounts

In this section I compare my account with three broad approaches found in the literature: (i) canonical teleosemantic accounts, which like my account appeal to an asymmetry in the way conditions explain successful behaviour; (ii) accounts based on decoupling at input and output; and (iii) accounts based on the system being able to detect when a directive representation has been satisfied (i.e. when a goal has been reached). I also reply to an objection that has been made to teleosemantic approaches.

Canonical teleosemantic accounts of the descriptive–directive distinction (sometimes referred to as 'direction of fit') appeal to the existence of a producer and a consumer of a representation (§1.4), something that is not available in my framework. Nevertheless, my distinction is closely related to the canonical teleosemantic treatment. According to teleosemantics, a representation R has directive content C when it has the function of causing a consumer system to produce outcome C (Millikan 1984, ch. 6). (It follows that R itself has the function of producing C, by that means.) A representation R has descriptive content when it is produced by a producer mechanism that has the function of producing R when condition C obtains, which in turn means that C's obtaining figures in an explanation of how behaviour of the consumer, prompted by R, led systematically to survival and reproduction.[2]

My definition takes from teleosemantics the core idea that directive contents are outputs which a system generates and which explain why a pattern of behaviour was stabilized; and that descriptive contents are other conditions whose obtaining explains how those outputs are produced and lead to stabilization. (Stabilization for me is not restricted to selection; and I add explanation of robustness.) My approach covers cases with multiple interacting internal components (so does not require a producer and a consumer). It also has a correlation or correspondence requirement at input (Chapter 4, Chapter 5). The obtaining of an exploitable relation (correlation or correspondence), in virtue of which the representations had systematic connections with external conditions, figures in the explanation of task functions—not just the fact that those conditions happened to obtain on occasions when behaviour had results that led to its stabilization. However, when dividing approaches to mode of representing into broad camps, my approach falls into the same family as canonical teleosemantics.

Marc Artiga argues that Millikan's account of directive content entails that all representations have directive content, which is implausible (Artiga 2014a). (Artiga uses this conclusion to argue in favour of an account according to which only descriptive content is attributed to simple representations: p. 552.) For there to be descriptive content in the first place, there has to be some range of behaviours which representation R is supposed to prompt its consumer to produce. That range may be very wide: $O_1 v O_2 v \ldots O_n$. Nevertheless, it follows, according to Artiga, that R has the directive content to produce $O_1 v O_2 v \ldots O_n$.

[2] In Millikan's terminology, C is a Normal condition in the most proximate Normal explanation of the behaviour of the consumer system prompted by R (Millikan 1984).

This kind of disjunctive condition is unlikely to arise as a directive content in my framework, since correlating with $O_1 v O_2 v \ldots O_n$ is a poor candidate for exploitable correlational information or for explanation, on usual approaches to causal explanation. First off, highly disjunctive properties are generally poor candidates to figure in nomological generalizations, hence in exploitable correlational information. Furthermore, think about candidate algorithms for performing S's input–output function. No such algorithm is likely to call for a stage of processing where such a nondescript fact is represented. In any event, the fact that a very general umbrella condition like $O_1 v O_2 v \ldots O_n$ obtains is unlikely to explain why a particular output F is robustly produced, nor to explain why it has been stabilized. My focus on how a collection of available internal states collectively explain task functions also means that there will generally be different contents for different vehicles (§6.2f), whereas Artiga's proposed content attribution results in very many representations having just the same (highly disjunctive) directive content. So, Artiga's objection does not apply to my account: a representation's having descriptive content does not entail that it has directive content.[3]

A second broad family of approaches identifies the descriptive–directive difference with an asymmetry in the coupling or decoupling that exists between a representation, on the one hand, and inputs and outputs, on the other. According to this approach, directive contents are still outputs that the system is supposed to produce, but there has to be a certain amount of decoupling between the representation and behaviour for it to count as a representation at all. So Price (2001) says directive content only arises when a representation has been selected to produce a particular outcome via a range of different behaviours in different circumstances (Price 2001, p. 141). It acts as a goal, on the basis of which the system selects, from a range of possibilities, a sequence of movements that is suitable for achieving that goal, performing a simple inference that also relies on descriptive representations about conditions in the environment. If the mediating behaviours are not themselves produced by a range of different means (if they are not robust outcome functions of the system, in our terminology), then they cannot be directively represented on Price's view. As well as lacking directive content, the motor programs driving these behaviours need not have descriptive content either. So, the important role of motor programs would be left out in the representational explanation of behaviour if we follow Price's definition.

Sterelny (2003) distinguishes between two kinds of decoupling (2003, pp. 30–40). Representations with 'response breadth' are not tightly functionally coupled to specific types of response (like Price's condition on having directive content). Representations that show 'robust tracking' are those which function by making use of a variety of cues so as to correlate with a feature of the environment. Sterelny argues that response breadth is a defeasible way of distinguishing descriptive from directive

[3] A similar reply is available to Millikan since directive contents derive from the 'most proximate Normal explanation' of behaviour, which will both count against disjunctive outputs figuring in the explanation, and count in favour of different representations in the same range having different contents.

representations: descriptive representations tend to be strongly decoupled from producing any specific type of response.

By making the converse point about directive representations, we would have an asymmetry in decoupling which differentiates between descriptive and directive representations. Kevin Zollman makes just this distinction, working within the sender–receiver models developed by Skyrms (Zollman 2011). Zollman differs from Skyrms who embraces the idea that all signals have both descriptive and directive content, content for Skyrms just being a function of the correlational information carried about inputs and outputs (Skyrms 2010). Zollman (2011) argues that an asymmetry between the correlational information carried about inputs and that carried about outputs must exist if a representation is to have only one direction of fit (2011, p. 163). He describes a signalling game in which such asymmetries arise. Descriptive representations are more tightly tied to inputs—they carry more information about world states—and directive representations are more tightly tied to outputs—they carry more information about acts.[4]

My distinction tends to go along with this kind of decoupling. Where R has directive but not descriptive content, production of outputs figures in the explanation of task function performance, so R will tend to carry stronger correlational information about outputs and to be decoupled from any specific input. The converse applies to a representation that carries descriptive and not directive content. It will tend to be more decoupled on the output side. I noted above that it would be a mistake to make these causal facts the basis for the distinction. Furthermore, doing so fails to capture the descriptive contents that are present in the kind of case mentioned in §7.2, where a motor program for movement B is relied on in generating further kinds of behaviour which make sense in the light of the fact that B is produced (the motor program thereby acquiring secondary descriptive content).

My view differs more markedly from a different kind of decoupling proposal, where decoupling is not just a causal matter, but depends on the kind of psychological processes that are performed on a representation. This idea stretches back to the way Lewis draws the distinction (Lewis 1969), and gets its intuitive support from the way hearers interpret and react to linguistic commands. Characteristic of a command is that the hearer doesn't have to deliberate about what to do (e.g. 'attack by sea'). The hearer can comply just by following the command. Assertions also have consequences for action (e.g. if 'the British are coming', there is probably something you should do about it), but the hearer has to deliberate about what to do in the light of that information. Huttegger (2007) models the distinction in this way by building the possibility of deliberation into his model.

[4] This is not Sterelny's distinction, since 'robust tracking' implies strong correlation with (distal) inputs, but it is compatible with Sterelny's view if we take his 'robust tracking' condition to be a requirement on being a cognitive internal state at all, rather than a tool for distinguishing between directions of fit.

Artiga suggests something similar when he says that decoupling is not just a causal matter, but that imperative representations are those that produce their behavioural results automatically (2014a, p. 558–9). That calls for a distinction between automatic and non-automatic use of a representation. The latter presumably involves deliberation, but as ordinarily understood, none of our case studies involve deliberation by the subject. There is processing involved in generating representations and in acting on them, but it is hard to see how a distinction between automatic and non-automatic processing can be applied. So, if 'not automatic' means something more than just 'decoupled', this is not a promising way to differentiate between descriptive and directive content in our cases.

A third general approach picks up on a different feature of a goal, which is that there is no point in pursuing it further once it has been brought about. On this view it is constitutive of having directive content about output C not only that the system is disposed to bring about C, but that it is sensitive to whether C obtains, so that behaviour prompted by the representation should cease once the system detects that C does obtain (Dickie 2015, p. 282).[5] Millikan makes this a criterion, not of having directive content at all, but of being a pure directive rather than a pushmi-pullyu representation (Millikan 2004).

That is a demanding constraint if it is taken to be a condition on having directive content at all, since it requires the system to have separate descriptive and directive representations about condition C, and to compare them. So, it would deny directive content in many of our simple cases, even when it is clear that it is the correlation with outputs that is explanatorily relevant, and whose coming to obtain (or not) explains the success or failure of behaviour. My approach is compatible with the intuition behind this view, however, because my distinction entails that representations which meet this demanding condition will have directive content. The toy model in Chapter 3 is an example (§3.6a and §4.1b). Where a comparator mechanism takes as input two vehicles and drives behaviour until they match, if the pattern of behaviour is such that one correlates strongly with the current state and the other with the output that tends to be produced, then the first will come out as descriptive and the second as directive (provided, as always, those correlations figure in an unmediated explanation of the system's performance of task functions). In short, there being a comparator mechanism is not a plausible requirement on the existence of directive content, but it is one way in which a system can come to have pure directives that are different from its pure descriptives. It is one way of making a system more sophisticated. The next section looks more closely at various kinds of cognitive sophistication that arise in directive systems.

[5] Smith (1987) is in the same spirit: it is constitutive of being a *desire* that p (having the directive mode of representing) that the state tends to endure in the face of the perception that not-p and dispose the subject in that state to bring it about that p (at p. 54).

7.5 Further Sophistication

(a) *More complex directive systems*

When we look at human beliefs and desires we see further layers of sophistication, which go beyond simply having separate descriptive and directive representations. This section briefly mentions four kinds.

In the human belief-desire system, orthodoxy holds that the same vehicle of content can be entertained with different attitudes. The vehicle of a belief can be redeployed to form a corresponding desire, and vice versa. That is not generally true in our case studies, but we did see two examples where it does arise: when place cells are used to guide action; and when the SINBAD network is used both to represent what is the case and to direct behaviour. We also saw, in the case of motor programs, that the correlational information carried by a directive representation could be made use of by other systems for its exploitable correlational information about outcomes so as to give it additional descriptive content.

The phenomenon of conditioned reinforcement is another example. In instrumental conditioning, obtaining a reward—that is detecting a certain kind of feedback—cements or promotes an internal configuration that encourages the disposition to produce the same kind of behaviour in same circumstances (Dretske 1988).[6] A stimulus that has never elicited a reward can itself become a reinforcing feedback if it has been repeatedly paired with a primary reinforcer. For example, if a rat has repeatedly observed a light paired with the delivery of food, then the light becomes a secondary reinforcer. The rat will then learn to act in ways that cause the light to go on, even if the light is not paired with food during that learning phase (Colwill and Rescorla 1988). The phenomenon of secondary reinforcement strongly suggests that a representation that descriptively represents that a condition C obtains can come to be a directive representation that functions to make the animal bring it about that C obtains.

Dickie's account of directive content (Dickie 2015) and Millikan's account of purely directive content (Millikan 2004) rely on a directive representation producing behaviour until the system detects that its satisfaction condition has come to obtain. It is usually assumed that this is achieved because the same vehicle is used for the descriptive and directive representation, making them readily comparable. On my account, keeping track of whether you have reached your goal is not required for having directive content, even for having purely directive content. Keeping track of satisfaction is an additional level of cognitive sophistication which, where it arises, does produce a descriptive–directive difference in the way that is usually assumed. The general-purpose redeployability arguably found in the belief-desire system is a further level of sophistication again.

One final level of sophistication is worth mentioning. Sometimes directive representations conflict: they concern conditions in the world that exclude one another, or the actions required to bring them about are different and cannot both be performed at

[6] See §4.2 and §8.2e.

once. Many organisms have a system for sorting between their directive representations to prioritize which ones to act on. That is not the same thing as being able to engage in practical reasoning: to reason from a directive representation (bring about C) and a conditional belief (C is likely if B) to a new directive representation (bring about B). (Arguably secondary reinforcement is a simple instance of that pattern.) The ability to sort amongst and prioritize goals is a further characteristic feature of the human belief-desire system. Although not constitutive of a representation's having directive content, its operation is a set of functional roles that is likely to imply that the vehicles so deployed have directive content. A characteristic way of doing that is for desires to have different strengths.[7] Relative strength can vary over time, for example through the vehicles being more or less active.[8] The choice of action then depends both on the relative strength of the agent's desire for q and the agent's assessment of how likely it is that they can bring about q.

Taking stock, there are at least four levels of sophistication involving directive representations. (There are doubtless others too.) First, there is having separate representations with purely descriptive and purely directive content, something not found in the simplest pushmi-pullyu systems. Secondly, there is the ability to keep track of when a directive representation has been satisfied, hence to compare descriptive and directive representations concerning the same worldly condition. Thirdly, there is the ability to redeploy the vehicle of a descriptive representation in a mode that gives it directive content concerning the same condition, and the converse. Fourthly, there is the ability to calculate over directive representations in order to prioritize which ones to carry out. I have argued that none of levels two, three, or four are needed in order to have representations at level one, pure directives and pure descriptives.

(b) *Another mode of representing*

Propositional attitudes admit of other modes of representing in addition to the descriptive and the directive, for example supposing. Something like supposing may be at work in one of our cases. When the rat navigation system is run offline to calculate the shortest route, place cell activation does not correlate with or represent where the animal is at that moment. Instead, activation represents somewhere it might be. Co-activation still represents the spatial relations between places, so when one place cell causes the activation of another offline, that represents that one location is near another, or that being at one location the animal could move directly to the other. So, the relation between vehicles is still a descriptive representation of spatial structure. The system is using that representation to do a kind of conditional reasoning: if you were at x, then you could get to y, and then to z, and so on.

[7] An agent engaged in planning or means-end reasoning has to have some motivating state, some directive representation. If it has two or more, and they call for control of the same effectors, then there must be facts about how likely each is to bring about its outcome in the presence of the others. That is their relative strength.

[8] We saw something similar with descriptive representations where the level of activation represented the probability that a particular world state obtains (§4.8).

Activation of an individual place cell in this process has neither descriptive nor directive content. One possible explanation for that is that it has unsaturated content, as we have seen (§6.3). Here I want to look at a second possible explanation, namely that it has content in a different mode of representing, a kind of suppositional or hypothetical content. Activation of a place cell offline concerns the same condition C as when it is activated online with descriptive content (Chapter 5) or directive content (§7.4 above), but it has a different overall content. It says something like *suppose you were at x*. That causes another place cell to fire, by virtue of the connection between vehicles. Instantiation of the co-activation relation represents, under the UE structural correspondence, that location y is near to location x. And the second place cell is also saying something hypothetical: *y would be nearby*. That is the conclusion of a little chain of reasoning: from *suppose you were at x*, and *y is near x*, to *y would be nearby*.

Now I have offered two possible explanations for why the activity of an individual place cell offline does not have a correctness condition: because it is unsaturated (§6.3) or because it has suppositional content (here). Which should we prefer?[9] It seems to me to count slightly against the unsaturated view that online place cell activity has a saturated content (*you are at x*). The suppositional view says that, when the vehicle is reused offline, the same saturated condition is in play, albeit in a different mode of representing (*suppose you were at x*). The unsaturated view calls for a switch from saturated content online to unsaturated content offline. On the other hand, the unsaturated proposal seems like a simpler account of the offline reasoning, since it just runs off straightforward descriptive representations like *location x is near location y*. I don't propose to resolve the issue. Either way, offline place cell activation introduces kind of sophistication which might, at first pass, be thought to be the preserve of propositional attitudes.

I won't speculate here on whether the suppositional content I have just described corresponds to the mode of representing of any propositional attitude state (e.g. supposing), or whether it can be captured properly by any linguistic term. I will rest with noting the interesting functional role of the representations involved in this offline reasoning.

7.6 Conclusion

This chapter draws the descriptive–directive distinction in a non-theory-neutral way: it shows how a distinction along these lines arises straightforwardly within the varitel framework. The account retains the virtues of the standard teleosemantic treatment of mode of representing, while being preferable to other existing theories, and also claiming some advantages of its own. Neither decoupling, nor the ability to keep track of goal satisfaction, are constitutive of having directive content. However, my account does imply that, where they exist, these features will give rise to a descriptive–directive difference as expected. The account can readily be applied to the case studies based on

[9] At the cost of some complexity, it is even possible to combine the views, so that an unsaturated constituent is used suppositionally, rather as we might say 'consider Nisha for a moment'.

UE information (Chapter 4) and UE structural correspondence (Chapter 5), with plausible consequences. There are several other kinds of sophistication which, while going along with a descriptive–directive difference, are not fundamental to it. Interestingly, the rat navigation case gives us a possible subpersonal example of a mode of representing that goes beyond the descriptive or the directive. Something like supposing may be involved when place cell activity is used offline in calculating shortest routes. In short, the varitel framework supports a useful way of understanding the descriptive–directive distinction.

8

How Content Explains

8.1 Introduction

This chapter offers some theoretical reflections on the accounts of content presented in previous chapters. First, in this section I briefly reiterate some of the distinctive features of my view.

One unusual feature of the book is that it devotes so much space to a series of detailed case studies. The aim of that was to understand how representation is used right across the cognitive sciences. Where it leads is pluralism. With two exploitable relations and a variety of task functions, my 'theory' of content is in fact a collection of different theories. Pluralism has been suggested before, but not until now worked up into a collection of detailed, mutually compatible accounts. My approach is unusual in focusing exclusively on subpersonal cases, and in the extent to which I'm interested in neural representation. Renouncing representation consumers is a new way to develop teleosemantics. We can get the benefits of representationalism—the explanatory benefits that flow from having vehicles of content—without consumers, and also without collapsing into an instrumentalist or ascriptionist view.

Being careful about the value of RTM led to the view that content arises from convergence between task functions, internal processes and exploitable relations: it arises when internal processing over vehicles standing in exploitable relations to the environment implements an algorithm for performing the organism's task functions. The idea of vehicles being processed in virtue of non-semantic properties in ways that respect their contents is of course not new; nor is the idea of exploitable relations (Godfrey-Smith 2006). However, the way varitel semantics puts these ideas together as the basis for content determination is distinctive. The focus on explaining the explanatory purchase of representational content is also a new emphasis, leading to an original proposal about why the world affords us the representational scheme of explanation—because of there being a natural cluster in which stabilizing processes go together with robust outcomes and internal mechanisms for producing them.

Section 8.2 returns to the question of content's explanatory purchase and shows that the accounts in Chapters 3–7 do deliver on the promissory note in Chapters 1 and 2. The varitel accounts allow us to see how the practice of representational explanation works and why content has an explanatory role to play. Section 8.3 looks at the causal efficacy of semantic properties, on the varitel view. Section 8.4 asks whether exploitable relations are doing any substantive work in the accounts, or whether they are dispensable in favour of an output-only approach to content. Section 8.5 asks how far varitel semantics extends, whether it applies to cases where content is un-explanatory, and whether it applies too widely. Section 8.6 remarks on the tight connection that exists between content determination and the circumstances in which a representational capacity develops, suggesting that this fits well with other issues in the literature. Section 8.7 goes through a list of clarifications and qualifications that couldn't easily be dealt with earlier. Section 8.8 draws out some epistemological consequences from our (metaphysical) accounts of content determination. Finally, section 8.9 suggests some ways that differences at the personal level may turn out to be relevant to content determination there.

8.2 How Content Explains

(a) *Explanatory traction in varitel semantics*

The varitel framework was motivated by the desideratum that we should be able to explain how content-based explanation works. The answer sketched in Chapter 2 started with the idea that contents have real vehicles because content explanation is partly concerned with explaining how a system manages to generate appropriate behaviour. Contents are externalist because the patterns of behaviour to be explained are world-involving: achieving distal effects in the world by reacting to distal objects and properties. The extrinsic properties that are relevant to explaining how the organism does that are exploitable relations that vehicles of content stand in to features of its environment. These externalist properties are suited to explaining how internal processing implements an algorithm for carrying out an organism's distal functions. That was all effectively a promissory note: if I can devise a theory of content that fits within the framework, then that should allow us to see how representational contents are suited to explaining behaviour.

Now that I have pinned my colours to the mast and set out a series of accounts of content, the time has come to assess whether the accounts deliver. Do they allow us to see how contents explain behaviour? In particular, do they throw light on the characteristic explanatory grammar of representational explanation (§2.2): that correct representation explains successful behaviour, and misrepresentation failure?

That explanatory grammar arises naturally from my accounts of content. Take the analogue magnitude system as an example (§4.6a). Consider a primate trained to choose between two sets of objects, being rewarded for selecting the more numerous collection. The training has used and tuned the animal's analogue magnitude subsystem, giving the animal a disposition to pick the thing in the world corresponding to whichever analogue magnitude register in its parietal cortex is more active. The training also gives rise to a standard of success and failure for the animal's actions in these contexts. When presented with two buckets containing collections of objects, picking the more numerous collection is a successful behaviour. Behaving in this way is a task function of the organism.

This task function not only constitutes a standard for success and failure of behaviour. It also specifies a mapping from distal situations (e.g. objects in buckets) to distal outcomes (e.g. picking up the fuller bucket), a mapping mediated by the animal. The monkey in its environment is disposed to instantiate this input–output mapping (at least approximately). There may be more than one algorithm that would generate this input–output performance, but there is a fact of the matter about which algorithm is at work inside the monkey. The algorithm is specified in world-involving terms: individuate the objects in one bucket, add up the total number of objects on that side, compare the total number of objects on each side, choose the collection with the largest number of objects. Internal processes inside the monkey are specified in intrinsic terms: patterns of neural firing here cause patterns of neural firing there cause... cause bodily movements. What makes it the case that this internal process implements an algorithm for performing the task is the correlational information carried by each component. Each component correlates with a distal fact called for by the algorithm (e.g. the number of objects on one side). Furthermore, the way internal processes operate over these vehicles implements the transitions called for by the algorithm. Thus, having content constituted by the convergence between exploitable relations and task functions, in the ways set out in previous chapters, both implies that there is a difference between successful and unsuccessful behaviour, and also makes for contents suited to explaining how an organism responds to distal facts in its environment so as to produce the distal outcomes which count as successes. Correlatively, misrepresentation by an internal component will explain unsuccessful behaviour.

I argued in Chapter 3 that this whole explanatory practice gets traction because of a deep fact about the world we live in. Things produced by natural selection tend to be disposed to produce outcomes robustly, because when an outcome is the target of selection, evolution can find ways for it to be achieved more robustly. Evolution's greatest robustness trick is the organism itself: a complex system that differentiates itself from its environment and continually maintains itself in a state that is out of equilibrium with the environment. Organisms produce the conditions needed for their own persistence, including by modifying their dispositions to achieve that end through learning. Furthermore, learning itself is a stabilizing process by which an individual can come to produce outcomes more robustly. So it is no accident that the biological realm is full of goal-directedness in the Aristotelian sense: robustly produced outcomes that have been stabilized by natural selection, learning, or contributing to the persistence of an organism.

One very common way—although by no means the only way—that organisms produce outcomes robustly is by having an internal mechanism that keeps track of aspects of the environment and thus implements an algorithm for performing the input–output mapping that has been the target of stabilization. That is, they do it by using representations. Nature has given us a widely implemented cluster: stabilization and robustness achieved by internal workings bearing exploitable relations. Representational explanations take advantage of the generalizations and inductions afforded by this natural cluster. Artefacts that humans have designed, like control mechanisms and computers, also often fall into the cluster, and for the same reasons. Tying representation to this cluster, rather than something more liberal, is the source of its inductive power: UE information and UE structural correspondence get explanatory purchase from the fact that their instantiation goes along with a cluster of other properties.

(b) Non-semantic causal description?

In addition, varitel semantics allows us to answer a familiar challenge to the status of representational explanation (§2.3). Isn't there an entirely non-semantic causal description of how any organism or system will react to inputs, undergo internal changes, and produce outputs? Realists about representational vehicles are committed to there being a non-semantic (vehicle-based) causal description at the same level as the semantic description.[1] If we can give a non-semantic causal description of the internal operation and outputs of a system moment-by-moment, what does representational content add?[2]

[1] I.e. at the same level of aggregation: semantic properties are properties of the very same objects (i.e. vehicles) as are found in a vehicle-based causal description of a system's operation.

[2] The non-semantic, vehicle-based description says how inputs to the system effect changes to internal vehicles, how those in turn influence further vehicles, and how that internal process eventuates in bodily movements. It is simplest to think of this as a complete causal description, saying how the system would react to any kind of influence on it. However, the syntactic description is itself a set of special science

Advocates of the explanatory force of representational explanation can point to all the successes of representation-based psychology—a huge body of work containing rich generalizations about representations in general, and especially about specific kinds of representation (motor programs, reward prediction errors, analogue magnitude representations, etc., etc.). But the vehicle-based challenge threatens to undermine the seemingly obvious explanatory traction of psychology by showing that it has no autonomy from a non-semantic element in its foundations. The aim of this subsection is not to catalogue the rich generalizations that give representational explanation its explanatory potency—we can turn to any psychology textbook for that. It is to show how varitel semantics answers the challenge. Varitel semantics has a feature which allows representationalism's commitment to non-semantic vehicles to be compatible with content having distinctive explanatory purchase.

To return to Ramsey's example (§2.2), consider the firing mechanism of a rifle (see Figure 2.2). Some theories of content imply that the displacement of the firing pin represents that the user's finger has been pulled back and instructs the cartridge to fire a bullet. If semantic contents were like that, then representational explanation would march exactly in step with a 'factorized' causal chain:

(i) The user's finger moves backwards.
(ii) The trigger moves backwards.
(iii) The firing pin shoots forwards.
(iv) The charge in the cartridge ignites.
(v) The bullet flies off at speed.

Steps (ii)–(iv) form a causal chain given in terms of intrinsic properties of the rifle. Step (i) is external to the rifle and causes (ii), which is a process intrinsic to the rifle. Step (iv) is also intrinsic to the rifle and causes step (v), which is an outcome external to the rifle. The causal chain is factorized into events in the external environment, on the one hand, and events intrinsic to the rifle, on the other. If movement of the firing pin were to carry semantic content, then there would be another explanation of the rifle's behaviour: movement of the user's finger leads to a representation being tokened with the content *the user's finger has been pulled back, fire a bullet*, which leads to a bullet being fired. That explanation marches exactly in step with the non-semantic explanation above. It is just a semantic relabelling of the process (i) → (iii) → (v).

Varitel accounts of content imply that semantic explanations of behaviour do not march exactly in step with a factorized causal explanation of how an organism

generalizations, and as such there will usually be things that can happen to the system that it overlooks, that appear as exceptions to its generalizations. For example, a stronger gravitational field might modulate the way a system performs actions, without that change being mediated by any differences in the vehicles involved in internal processing. A different kind of example is where an unusual influence on internal vehicles—e.g. through transcranial magnetic stimulation (TMS)—makes changes to an intermediate step of internal processing in a way that bypasses changes to upstream vehicles and so shows up as an exception to the causal transitions described by the algorithm.

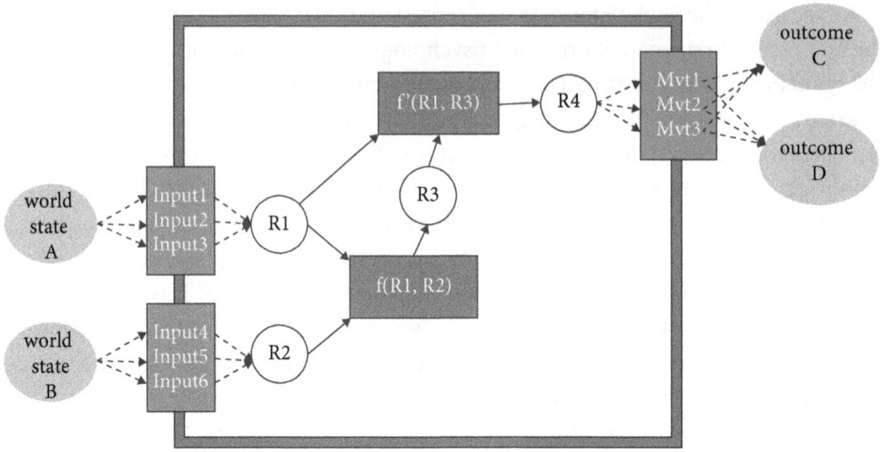

Figure 8.1 A schematic depiction of bridging at input and output.

responds to proximal inputs with bodily movements. Task functions are robust outcomes, so the same outcome is produced in response to a range of different proximal inputs. That means that vehicles of content will enter into generalizations that 'bridge' across multiple proximal conditions and involve distal states of affairs (see Figure 8.1).[3] Connections like the one between steps (i) and (iii) above, rather than being mediated by a single proximal input (ii), will be mediated by a range of different proximal inputs (ii*), (ii**), etc. There is a distal exploitable relation without a matching proximal exploitable relation. The representational explanation does not then simply march in step with the factorized explanation. Causal steps which show up as different in the factorized explanation are unified in the representational explanation. The representational explanation is picking up on patterns in vehicle–world relations that the factorized explanation would miss. The same is often true at output since outcomes that count as task functions tend to be produced by the organism in a variety of different ways in different circumstances; that is, via a variety of different bodily movements (§3.3 and §3.6). Those are further patterns, captured by the representational explanation, that a factorized explanation would miss.

This bridging means that there are real patterns in organism–world relations—in relations of internal states to distal causes and outcomes—that are treated disjunctively in the factorized explanation. The effect of past processes of stabilization has been to key the organism into the world so that generalizations do not just concern how proximal causes affect the organism and how the organism affects its immediate proximal environment. A purely factorized explanation would miss the distal patterns. Dennett argued that belief-desire explanation picks up on real patterns in the way

[3] This is not a novel feature of varitel semantics; e.g. a requirement for robust tracking (Sterelny 1995) or constancy mechanisms (Burge 2010) has the same effect.

agents interact with the world (Dennett 1991). Not only would prediction of agents' behaviour be impossible in practice if we treated them as collections of molecules interacting with other molecules in the environment, but also such a low-level physical description would fail to capture real patterns that exist—that are an interpreter-independent feature of a world full of agents.[4] My approach to content similarly shows how content-based explanation takes advantage of real patterns in the world, patterns that exist because of regularities in biological and physical processes, and which exist irrespective of the existence of observers to notice them.[5]

In the absence of a robust outcome function this argument does not get off the ground. Then a factorized explanation may march exactly in step with the putative representational explanation. We saw that in the case of the rifle. That is not a case where representational content would give any additional explanatory purchase, and indeed varitel semantics does not imply that movement of the rifle firing pin would carry representational content. The same point applies to the magnetotactic bacteria. As the case is standardly described (Dretske 1986, Millikan 1989, Cummins et al. 2006), moving in the direction of oxygen-free water is a stabilized function of the bacterium's behaviour, but not a robust outcome function.[6] Our accounts do not imply that the bacterium or its magnetosome carries representational content. Nor is it a case where content would afford a better explanation than a non-semantic causal description (one involving functions but not representational contents).

We can see bridging at work in our case studies. Consider again a monkey that deploys its analogue magnitude system to choose the more numerous set of objects in a range of different circumstances. At input, analogue magnitude registers in the parietal lobe correlate with the numerosity of distal collections of objects, a correlation that is mediated by a variety of proximal inputs: many different kinds of visual patterns, auditory patterns, etc. At output, the action of reaching out to touch or grab the more numerous collection is mediated by a variety of different motor outputs in different circumstances. So, there is bridging at input and at output. The rat's system for spatial navigation, using the hippocampus and other areas, also shows bridging. The input sensitivity of place cells bridges across multiple patterns of sensory input (e.g. visual input from different orientations). Behavioural output also exemplifies

[4] The question of the observer-dependence or observer-independence of these patterns is orthogonal to another feature of Dennett's view, the fact that it does not commit to there being vehicles of content ('ascriptionism', §1.3). RTM's commitment to vehicles could be combined with the view that the contents represented by those vehicles are observer-dependent.

[5] To be part of a real pattern, a content property underpinned by the bridged relation needs to show up in generalizations connecting it to others (e.g. in psychological theories). I am not committed to Dennett's particular way of theorizing real patterns in terms of Kolmogorov complexity. I am just relying on the idea that the generalizations are based on the underlying system mostly being organized in a particular way, and on that being a non-observer-dependent feature of the world (contra for example perspectivalism; Craver 2013). Many real patterns do compress information in the sense that they only allow a 'probabilistic recovery of the underlying system' (Ladyman 2017, p. 153).

[6] It also probably lacks sufficient internal structure to count as implementing an algorithm by which it achieves its functions. Indeed, dead bacteria continue to rotate into alignment with a magnetic field (Cummins et al. 2006, Schulte 2015).

bridging. The rat can reach a baited or rewarded location from a range of different starting positions by a range of different routes, relying on structural correspondence in the way described in §5.3.

If we were just to look at the rat's limb movements taken in isolation, without considering how those bodily movements produced locomotion or where the animal was in space, we would see a variety of nearly uninterpretable movements of different limbs with different speeds and in different directions at different times. It would be like watching the bodily movements of a teenager playing a video game on a smartphone, but without being able to see the screen. The two thumbs move rapidly in seemingly arbitrary ways, with no apparent pattern. Only once the relation of those movements to what is happening on the screen comes into view do they become interpretable. The real patterns are not found in when the thumb moves, but in what happens to the character on the screen, and how that relates to movements made by the game-player, and to her intentions. There are real patterns in what is happening in the player's environment. The thumb movements act as 'bridging' causal intermediates. Similarly, there are very clear patterns in the rat's behaviour if we consider it in relation to its environment. Those patterns are mediated by and generalize—bridge—across a diversity of bodily movements output by the rat.

The content-based generalizations in psychological theories link representation with representation, often of specific types, representation with neural substrate, and representation with world. These are by no means all cases of bridging. What bridging does is to show how content-based explanation can break free from non-semantic vehicle-based explanation, allowing the rich and detailed theories of psychology and cognitive neuroscience to have their own explanatory purchase.

Bridging exemplifies one explanatory virtue, generality. It groups together things that would otherwise be classified as different. But the previous subsection (§8.2a) argued for a seemingly contradictory virtue, specificity. An account that made content very liberal would be problematic and an advantage of varitel semantics was that its contents only arise when a special cluster of properties occurs (which it often does, for natural reasons). There is in fact no contradiction because the advantage claimed for the cluster is its inductive potential. Finding UE information or UE structural correspondence implies many other things about the system in question. In fact, bridging relies on inductive power as well. Content is based on a bridged relation but shows up in generalizations with other properties. With explanation, there is always a balance between properties that apply widely enough to support good generalizations and properties that are specific enough that they support rich inductions. Our accounts show how content properties strike a balance that gives them genuine explanatory traction.

(c) Doing without talk of representation

Another kind of challenge is faced by any theory of content which offers non-semantic, non-mental, non-normative conditions under which representational content arises. Suppose it is granted that correlation, correspondence, and function come together in the special way I have claimed, so that their convergence affords generalizations and inductions. However, the objection runs, we can recognize all that without ever mentioning representation or content.[7] Why can't we do all the explanatory work directly in terms of correlation, correspondence, and function? Indeed, the accounts of content set out above give us precisely the tools we need in order to do just that.

There are two versions of this challenge, which should be answered in different ways. The first rival to content-based explanation only helps itself to the resources that appear in my definitions—correlation, correspondence, robustness, stabilization, etc.—and does not advert to the fact that these properties tend to come together in the packages I have pointed to. Instead of contents, they offer explanations directly in terms of the properties in the explanatory base. They replace content-based generalization with much more fine-grained explanations. The trouble with these views is their complexity. More complex properties are generally less good candidates for explanation. Furthermore, it is not clear why relational properties like correlating or producing outcomes robustly get any explanatory purchase, if the motivation I have offered in terms of a natural clustering of properties is absent. Such a rival would also miss the inductive potential that exists—the fact that robustness and stabilization do tend to converge (so as to constitute task functions) and the fact that internal workings, correlation, correspondence, and task function do tend to converge in regular ways (which is what I say constitutes content).

The second version of the challenge recognizes the clusters set out in our accounts of content, but objects to these being identified as representations. We can do all the same explanatory work by recognizing that there are these real clusters and making generalizations and performing inductions over instances of them. This supposed challenge is not really a challenge at all, because it concedes everything we need, leaving only a dispute about the appropriateness of the label 'representation'. A distinctive style of representation-based explanation that our theories of content need to explain is that correct representation explains successful behaviour and misrepresentation explains failure; that is a form of explanation where the obtaining or otherwise of facts that are distal to the organism or system make a difference to explaining its behaviour. Explanations like that are part of what made representational content puzzling, even mysterious. Recognizing that the clusters I point to are real and important features of the natural world, and accepting that they underpin world-involving explanations of this kind, just is to accept that properties of the kind I have characterized do exist, and do explain behaviour in the way I have claimed.

(d) Other views about the explanatory purchase of content

Now that I have laid out my view about the explanatory purchase of representational content, I will compare it briefly to some other views in the literature. William Ramsey

[7] E.g. Hutto and Satne (2015) with respect to some forms of intentionality; Egan (2014) with respect to cognitive content.

identifies two broad ways in which representational properties have been argued to earn their explanatory keep (Ramsey 1997, p. 37). First, they may have heuristic value in allowing us to think about a system in a useful way. They can play that role despite having no causal relevance to how the system operates.[8] Some adopt this view about the computational theory of mind: syntax does all the causal work, but semantics allows us to see why a system's syntactic processes are suitable for performing certain computations. The second option is that contents are causally explanatory of behaviour; for example, because they are a structuring cause of dispositions to behaviour, as argued by Dretske (1988). (A further view is that representational content does not earn its explanatory keep at all, and so should be abandoned, e.g. Stich 1983.)

Frances Egan is in the first camp (Egan 2014). Content for Egan (her 'cognitive content') is useful because it allows the theorist to understand how a computational system can perform a cognitive task; for example, the task of seeing what is in the nearby environment. Different contents will be useful for explaining how the same computational system performs different cognitive tasks. In each case, content is just a gloss, useful to the theorist. Oron Shagrir falls in the same camp, but with a more realist take on content (Shagrir 2006). For him too, theorists adopt the representational approach in order to explain semantic tasks performed by a system. These views are similar in one way to Tyler Burge's theory, since Burge takes the target of representational explanation to be the capacity for perception and the computations and transformations involved in perception (Burge 2010). In all these cases the task that calls for explanation is already stated in semantic terms. Contentful states enable us to understand how an organism can perform a cognitive, semantic task.

Dretske (1988) exemplifies the second camp, since he argues that contents are causally explanatory. Content arises when an internal state R has been recruited as a cause of behavioural output M in virtue of the fact that R carries information about[9] condition C. The fact that information has converged with learning in the past—the fact encapsulated by the existence of content—causally explains why the organism is configured as it is today, and hence forms part of an explanation of why it produces behaviour M on a particular occasion.[10]

Another way that contents can have causal relevance arises specifically for conceptual contents. Possessing concepts can explain why certain capacities of an organism are systematically related. The compositionality of concepts is then the source of a special

[8] That is not of course to deny that non-semantic properties of vehicles of content can be causally relevant. Causal relevance: an apple on the greengrocer's scales causes the spring to extend. Here the apple's mass is causally relevant and its colour is not.

[9] Dretske says 'indicate', which is a more restrictive species of correlational information. Godfrey-Smith (1992) objects that natural selection does not require indication but often makes use of weaker kinds of correlational information about adaptively relevant facts.

[10] Those who point to the relevance of externalist properties to explaining the behaviour of an organism in its environment—e.g. to sorting behaviour (Davies 1991) or action guidance (Peacocke 1993)—are also arguably in the causal relevance camp, although theorists who object to a causal role for externalist properties will see these as being cases where content has merely heuristic value.

explanatory value of representational locutions, because it allows us to explain the systematicity of cognitive capacities, although here too non-semantic vehicle properties are rivals for causal relevance (Camp 2009, Fodor 1987b).

Varitel semantics partakes of elements of both camps. Within the second camp, like Dretske I rely on considerations about why the system has been configured the way it is and behaves the way it does.[11] In varitel semantics content arises when an organism has a disposition to produce certain outcomes because exploitable relations converge with stabilizing processes that have operated on those outcomes in the past. Ramsey (2007, pp. 132–40) objects to Dretske's theory on the basis that it over-generates: not all cases where indicator properties are a structuring cause of behavioural outputs should count as representational. My accounts have stronger requirements than Dretske's and so avoid this liberality objection. Nevertheless, part of the explanatory power of content, on my view, traces to the kind of causal process identified by Dretske.

My accounts share with the first camp the view that contents are useful because they allow us to see why an organism's internal workings are suited to performing certain tasks. Unlike Egan and Shagrir, I characterize those tasks in non-semantic terms in the first instance. Mine are not cognitive tasks but mappings from worldly conditions to distal outputs (outputs which qualify as task functions). But like Shagrir, contents for me are partly a matter of how an organism can perform the computations needed to produce appropriate outputs in appropriate circumstances. I reject Egan's contention that contents are merely a theorist's gloss, with a different gloss being appropriate when a system is located in different contexts. The context in which a system is operating is a property of the system, just as much as its intrinsic properties are, and I take that context to have a content-determining rather than merely a pragmatic role.

Since concepts have not been part of our investigation, I have said little about the role of representations in explaining systematicity. Nevertheless, we saw in §6.3 that the representations in some of our case studies do have semantically significant constituent structure. Even without semantically significant structure at the level of representational vehicles, the division of an organism's internal workings into a series of algorithmic steps has some of the flavour of systematicity (§5.7a, §6.3).[12] In both cases, facts about vehicles and how they interact explain certain systematic patterns in the organism's behaviour. The point I have laboured about vehicle realism and internal workings (§1.3, §2.5, §3.2, §8.2a) is in fact a generalization of others' observations about systematicity of structured representations and the explanatory purchase of content (Fodor 1975, 1987b; Fodor and Pylyshyn 1988; Camp 2009).

In short, varitel semantics can adopt sound arguments about explanatory purchase drawn from both of the broad camps identified by Ramsey (1997, p. 37).

[11] Dretske argues that this form of explanation of behaviour is unavailable when it is natural selection in ancestors, rather than learning, which explains why the organism acts on R to produce M as output (Dretske 1988, p. 94; 1991, pp. 206–7). My framework covers both cases. Godfrey-Smith (1992, pp. 294–6) argues that natural selection should be assimilated to Dretske's scheme of explanation.

[12] But recall that 'computational structure' is not a case of structural representation: §5.7a.

8.3 Causal Efficacy of Semantic Properties

The previous section showed why representational explanations are partly autonomous from purely vehicle-based explanations, and hence why semantic properties can afford us some additional purchase in explaining behaviour. But are content properties causally efficacious, or simply explanatorily relevant? Jackson and Pettit distinguish between process explanation and program explanation (Jackson and Pettit 1988, 1990). They argue that the properties cited in a program explanation can be explanatorily relevant without being causally efficacious—when it is the properties cited in the process explanation that do the real causing. For example, the squareness of a wooden peg explains why it won't fit into a round hole of the same surface area. However, the causally efficacious properties, it is argued, do not involve squareness or roundness, but other physical properties of the material of the peg and hole.

Jackson and Pettit introduce their distinction in order to save the explanatory relevance of broad contents, and it is equally applicable to our case. The vehicle-based explanation of behaviour may be telling us where the real causal work is going on, but nevertheless content properties can be explanatory. This casts varitel accounts of content as picking up on opportunities for program explanation. When the explanandum occurs in a range of cases, and so is more general than any specific causal process, a program explanation tells us that what matters for achieving a result is that some relational state of affairs obtains, irrespective of which particular state of affairs causes it to obtain. Program explanations 'tell us about the range of states that do or would produce the result without telling us which state in fact did the job' (Jackson and Pettit 1988, p. 396).

If contents are fixed in the way I have claimed, then the argument in the previous section shows why semantic properties figure in program explanations, hence are explanatory of behaviour. That they are not causally efficacious is a further claim. One challenge is faced by all special science properties, the challenge that the 'real' causal work is being done at some more fundamental level. But if we thought that the real causal work was captured by the factorized, vehicle-involving explanations mentioned above, then we would face a further challenge: vehicles are realized by physical properties (e.g. neural firings). Surely the vehicle-based description is only a program explanation, with the causal work really going on at a more fundamental neurophysiological level? But, of course, the same move arises there, with a molecular, chemical, and electrical explanation threatening to displace the causal efficacy of neural depolarizations. The descent continues, until we reach the level, if there is one, of the most fundamental physics (where, incidentally, it is arguable that causation does not figure at all).

An alternative is that causation—'real' causal efficacy—is found at more than one level in this hierarchy. There may be no good reason to think that causal processes that are connected by relations of constitution should exclude each other (Bennett 2003). Thus, even if it is granted that some patterns of explanation deal in explanatory relevance rather than causal efficacy, it would be hasty to conclude that content-based explanations are not a locus of causal efficacy.

If causal efficacy is a tenable position for some special science properties, there are further obstacles to making that case for content properties. Contents are partly historically based, and are partly tied to the kinds of effects they are called on to explain (Shea 2007b). I have argued that these are not obstacles to explanatory purchase, but it is a further step to show that these features are compatible with the causal efficacy of content. We would have to see that the world-involving generalizations that contents are involved in support counterfactuals and interventions in the right way, and possibly that they count as figuring in a causal process (on process-involving views of causation). We would also need to establish that the relation between content properties and the vehicle-based story does not generate causal exclusion, especially in the light of the fact that some of the representation-to-representation transitions (or inferences) featuring in the semantic-level explanation have an exact parallel in the syntactic explanation. Nevertheless, it could be that the forms of generalization, counterfactual-dependence, and underlying process that embed these transitions in world-involving patterns amount to set of genuinely causal relations.

These are large issues, well beyond the scope of this book. For now I want to remain neutral about whether contents are causally efficacious. Varitel semantics allows us to see why semantic properties figure in program explanations, so I rest with the positive claim that content properties are at least explanatorily relevant.

8.4 Why Require Exploitable Relations?

My accounts require that representations bear exploitable relations to the things they represent. More carefully, exploitable relations have to be in place when behaviour is stabilized (when the stabilizing process, which makes it the case that the system has a function that is partly constitutive of content, is operative). The leading teleosemantic theories of Millikan and Papineau eschew any requirement of this kind. Contents are an output-only matter, determined by the functions of behaviour prompted by a representation, and by evolutionarily normal conditions for performing those functions. Are exploitable relations dispensable? It is not in dispute that exploitable relations will normally be in place during stabilization. For example, dances of incoming honeybees would have correlated with nectar location at the time of selection. But do correlations need to figure in the content-constituting story?

My account agrees with teleosemantics that functions are an essential part of the story. They furnish an explanandum to which representational explanation is directed, namely successfully producing distal effects in the environment, or failing to do so. Teleosemantics says that contents are a matter of functions and conditions under which functions will be successfully performed. Varitel semantics goes further. It says that content is partly a matter of explaining how a system achieves its functions—of how its internal workings implement an algorithm for performing its functions. That is why exploitable relations get into the picture.

Without exploitable relations, representational contents are effectively a way of typing internal states by patterns in the outputs they (historically) produce. Adding exploitable relations on the input side furnishes a reason why, when the representation is tokened, it is likely that its correctness condition obtains.[13] That will be true at the time of stabilization and, to the extent that things haven't changed, will also be true of current behaviour. That is, our accounts do not just produce correctness conditions that are a typing of behaviour. They give us a reason to think that those correctness conditions obtain—a stronger reason than teleosemantics is committed to.

That difference gives varitel accounts more predictive power: we can use content to predict what the system will do because, if the environment is stable in relevant respects, representations will continue to correlate or structurally correspond with the world, giving us stronger predictions about the distal results that are likely to flow from outputs produced by the system. That difference is not profound, however, because teleosemantics can appeal to something very similar: an empirical generalization that, where there are contents of the teleosemantic sort, exploitable relations are usually in place. So the predictive advantage of varitel semantics only arises in somewhat exotic cases where teleosemantics would ascribe contents to states that are generated at random (Shea 2007b, pp. 427–30).

The advantage of varitel semantics is more substantial when we come to explanation. My accounts show why contents are suited to explaining how systems perform their functions—they do so by making use of exploitable relations carried by their components. Teleosemantics is more in the business of typing representations by the behaviour they produce than in the business of explaining how a system produces behaviour.[14]

8.5 Ambit of Varitel Semantics

(a) *Representation only if content is explanatory?*

Is content in the eye of the beholder? Does the question of whether a system has representational properties depend upon whether it is useful for some observer to treat it as such? Not according to my accounts of content. Each account of content that falls within the varitel framework is given in terms of correlation, structural correspondence, robust outcomes, stabilization, and so forth. Why is that complex property of interest? Because it arises in the world for natural reasons and, where it arises, it generally

[13] It is agreed on all sides that representations produce correlations on the output side at the time of selection/stabilization: correlations with the distal effects they produce. This is a form of exploitable relation. However, that is not enough for there to be content according to my accounts. Exploitable relations at input also have to be in place: see §4.2a.

[14] Godfrey-Smith (1996, pp. 171–95) argues that teleosemantics makes representational explanation of behaviour akin to explaining why a sleeping pill put a person to sleep by citing its dormitive virtue. In Shea (2007b) I argue that, while dormitive virtue explanations are not empty, adding an exploitable relation requirement—in that case a correlation requirement—makes the representational explanation more substantial.

affords distinctive kinds of generalization and induction. Saying these properties allow us to engage in a distinctive form of explanation does not imply that the existence of these properties is relative to the existence of an observer, able and willing to go in for explanations of this kind. The existence of these patterns is an observer-independent fact, as is the fact that some properties explain others (§4.2b).

Since contents are not constituted by the explanatory practices of an observer, there is no requirement that contents should afford useful explanations in every case. Contents constructed in the ways I have set out are suited to getting explanatory purchase in many cases, but may not do so across the board. Consider for example a thermostat, one that is only slightly more sophisticated than the philosopher's standard example. At input, it has two ways of gauging room temperature, a sensor for levels of sunlight and a sensor for thermal expansion. At output it controls the temperature through operating a radiator valve and an external air vent. Its capacity to keep a room at a set temperature is slightly better than a normal single input thermostat. It predicts the warming effect of sunlight and so smooths out some of the bumps in temperature that a standard thermostat would produce. But it is only slightly better. Arguably, holding the room temperature constant is a robust outcome produced by thermostat. Deliberate design may constitute that outcome as a task function. And it is a task function achieved by (simple) internal workings that make use of exploitable relations between internal states and distal states of the environment (there is some bridging).

However, this is a case where we would get little or no additional explanatory benefit from the representational explanation than we get from a non-semantic causal explanation of how proximal inputs generate proximal outputs, and how that affects the temperature of the room. It is a case where there is representational content but not one where representational explanations are substantially better than non-representational or vehicle-based explanations of behaviour.

Part of the reason why representational contents vary in their explanatory purchase is that the features which give them explanatory bite come in degrees. A system can produce outputs more or less robustly: across wider or narrower ranges of proximal inputs, and by more or fewer different proximal bodily movements. Furthermore, task functions can arise from a range of different stabilizing processes: natural selection, feedback-based learning, and simple contribution to persistence of a biological organism. In paradigm cases all three stabilizing processes will be at work and will be pulling in the same direction. In less paradigmatic cases they may come apart, so there can be different contents constituted relative to different task functions. In marginal cases, or in thought experiments like swampman, maybe only one of these processes comes into play; for example, contribution to persistence on its own. Such cases will have representational content if they meet one of the conditions above. However, in these more peripheral cases representational content may well have less explanatory purchase than in paradigm cases.

Task functions constituted only by design, as in the thermostat example above, may also give rise to contents that are less explanatorily useful, depending on the amount of robustness and internal vehicle-based complexity that has been built into the system. On the other hand, some designed artefacts like a sophisticated computer-guided missile, with a high degree of internal complexity and the capacity to produce outputs very robustly, may be cases where the representational description is practically indispensable for explaining behaviour, in roughly the way Dennett claimed beliefs and desires are indispensable in practice for explaining human behaviour.

In short, representational contents are not in the eye of the beholder. They are constituted by the coming together of the properties mentioned in the conditions set out in previous chapters. The extent to which the representational explanations they afford are useful or practically indispensable will vary with the facts of the case.

Finally, I briefly assess whether the accounts of content put forward here are pragmatist. This label is sometimes associated with the claim that the brain exists to guide action, that it is therefore not in the business of building models of the world (Barrett 2011), and that we should not therefore expect representation to play a central role in the cognitive sciences (Anderson and Chemero 2016). Varitel semantics largely agrees with—indeed is based on—a version of the first claim. But it strongly rejects the second and third. The brain forms representations, and builds models of the world, and does so in the service of guiding action. Representational content in our case studies depends ineliminably on action, its functional significance, and the role of representations in guiding action. So varitel semantics has contents that are pragmatist in the sense that their content is action-derived, while rejecting the anti-representationalist conclusion that is drawn by those who claim that enactive engagement with the world displaces representational content (Hutto and Satne 2015).

The regularity with which I use the word 'explanation' suggests pragmatism of another kind; my emphasis on the observer-independence of content properties suggests otherwise. Simon Blackburn offers a useful characterization of what pragmatists are often up to (Blackburn 2010). The pragmatist aims to explain why we go in for a certain kind of discourse. I am certainly doing that. A major explanandum for my project is the pattern of representational explanation found in the cognitive sciences. I can be seen as offering a (realist) explanation of why psychologists and cognitive neuroscientists engage in that kind of discourse and think about organisms and systems in that way.

What divides my approach from pragmatism is the kind of explanation I offer. An explanation is pragmatist, according to Blackburn, when it eschews any use of the referring expressions of the discourse but proceeds by talking in different terms about what is done by the discourse; for example, by showing that it serves a particular function (Blackburn 2010, pp. 1–2). By contrast I argue that, in order to explain behaviour in the kinds of cases I am interested in, we do have to refer to representations (as real vehicles) and their contents. What I have offered is a meta-semantic metaphysical account of what various theoretical terms in use in cognitive science refer to—terms in the family of representation, semantic information, content, correctness condition, satisfaction condition, and so on. In explaining how this discourse works I use the terms representation and content as referring expressions, so my accounts of content are not pragmatist by Blackburn's lights.

(b) Are any cases excluded?

Does varitel semantics imply that every natural system is processing internal representations? Won't any system designed by natural selection, learning, or human design end up having internal states that count as representations? We have already seen that magnetotactic bacteria do not have representations (§8.2b). Similarly, the 'two component' signalling processes that are ubiquitous in bacteria (Lyon 2017) will be excluded when they depend on detecting just a single proximal cue.[15] Varitel semantics does extend readily to non-psychological cases, but there are principled reasons why many cases are excluded.

Consider the way plant roots follow local concentration gradients so as to move towards water (Takahashi 1997). Is the root (or the plant) representing the direction of water? No, because here the story about stabilization does not involve distal facts, but just how the root reacts to the proximal availability of water. Or consider a germinating seed that uses gravity to grow upwards towards the surface of the soil. The adaptively relevant fact here is distal, the availability of sunlight. Must it not have an internal representation of the direction of the sun? Must a plant that rotates in the direction of solar radiation have an internal representation of the direction of the sun? These input–output mappings, as described, are not the basis of task functions because they are not robust outcome functions. The output is modulated just by intrinsic properties of the sensory input. Distal features of the environment are adaptively important, but there is no task function (stabilized function + robust outcome function) that concerns distal features of the environment, and no mechanism that 'bridges' across multiple sensory inputs in correlating with a distal feature of the environment.[16] The representational content that would be attributed by standard teleosemantic accounts marches exactly in step with a factorized explanation of the plant's behaviour. These are not representing systems according to varitel semantics.

Which is not to say that plants never go in for representation, let alone that representational content is restricted to psychological systems. Consider a plant that opens its flowers in the day and closes them at night. Suppose it just relies on changes in temperature, which alter internal biochemical processes. The opening and closing behaviour is produced in response to only one input, and so would not be a task function. It would be an evolutionary function of the plant, but would lack the robustness to be a task function. Now supplement the case slightly, making it more biologically realistic, so that the plant is also sensitive to light levels, giving it a second way of detecting

[15] Even if the mechanism is very complex, as described in Hsieh and Wanner (2010).

[16] It is the absence of a distal-involving task function that stops representation arising. It is not a requirement on being UE information that the correlation should concern something distal (§8.7). (However, because of the requirement for a distal-involving task function, some items of UE information in the system need to concern distal features of the environment.)

that evening has arrived. Then the plant has two ways of detecting that it is evening, and so the flower-closing behaviour would be a (very simple) task function of the plant. Internal processes in the plant could then well be representations with descriptive content about the time of day and with directive content telling it when to open and close its flowers.

But surely robustness is not unusual, but an absolutely ubiquitous feature of biological organisms? Cells have robust metabolic networks (Krasensky and Jonak 2012). Cells are even able to explore and sample new possible metabolic networks when they are put under conditions of severe stress (Szalay et al. 2007). During cell development the spindle microtubules that structure the cell grow in the right places robustly—they do so because of a process of selection in which many spindles are started and only those that reach their targets are preserved (Kirschner and Gerhart 1998, pp. 8422–3). Both metabolism and cell development produce outcomes robustly, and do so as a result of stabilizing processes that operate on both the phylogenetic and the ontogenetic timescale. Surely these are paradigm cases of robust outcome functions? Not as I have been using the term. Our target is cases where the same outcome is produced in response to different external inputs and is robust in the face of changes in the distal circumstances in which it is produced. These cellular and metabolic examples are cases of robust internal processes. They do show adaptive responses to things happening at the cell surface, such as damage to the cell wall, but the functions there are described in terms of intrinsic properties of the cell and changes happening to it. These are not cases of the kinds of functions that can ground representational content, which is what my definition of task functions was designed to capture.

There is no reason why hormonal signals should be excluded from being representations within the varitel framework. Sometimes their operation will be explicable in purely intrinsic functional terms, but in many cases hormones are involved in tracking adaptively relevant distal facts by multiple routes (e.g. about conspecifics), part of the basis of task functions. Another set of cases is found in the immune system. It has complex mechanisms for detecting threats and responding adaptively, so it is very likely to help the organism perform task functions, and it would not be at all surprising if representations were involved in carrying out those task functions. In short, cases of internal subpersonal representation in organisms extend well beyond the psychological. Furthermore, these are cases where representation gets explanatory purchase: world-involving contents allow us to see how the hormonal system or immune system enables the organism to achieve certain distal outcomes in its environment.

Another way that subpersonal representations may arise internally is when there are task functions of systems that are smaller than the whole organism. For example, an individual cell is likely to have task functions, as may a larger unit like the immune signalling system. That system may have robust outcome functions concerning facts about other parts of the body distal to it. For example, an individual cell may have multiple ways of detecting the overall physiological state of the organism it is in (e.g. stressed vs. non-stressed) and responding appropriately. If so, the cell could have a task function, one that could be representationally mediated. By this route some processes going on within an organism might count as task functions, functions for systems that are smaller than the organism as a whole. Caution is needed here, though. Not all evolutionary functions will count. To be a task function, an output of a system needs to have been the target of a stabilizing process operating on that system as such. Learning-type processes within a cell count. The process of generate-and-test used by the immune system may also count. But outputs of an internal system that are stabilized only because of the way they contribute to a stabilizing process operating at the level of the whole organism do not thereby count as task functions of the internal system.

Hormonal signalling may be like that, its functions deriving from task functions of the organism in which it operates. It may not be sufficiently distinct to count as a system in its own right; and if it does, there may be no stabilizing process acting at the level of the hormonal system as such, independently of those operating at the level of the whole organism. The same may apply to the brain. There are lots of processes of selective retention going on in the brain, of course, but these play a role in whole-organism stabilizing processes (various kinds of learning), which counts against them having an intra-organism task function of their own.

Varitel semantics is more demanding than theories of content that are based just on evolutionary functions or just on correlational information. But that is not because it is designed to capture only a class of cases which seem, a priori, to count as calling for representations. So, it is no kind of desideratum that only psychological systems should be capable of forming representations. Instead, we took psychological systems as studied by cognitive science as a paradigm and aimed to account for what gives representational contents their explanatory purchase in those cases. We then find that that distinctive explanatory scheme extends more widely (§6.5b). For example, it includes the kinds of internal signalling in plants we've just discussed. It also covers many cases of between-organism animal signalling. Both the vervet's alarm call and the honeybee's nectar dance are cases where the representation producer integrates multiple cues in order to produce the signal. So, output behaviour is a task function, both in the monkeys and the honeybees.

When we come to the most familiar examples of representations in psychological systems, namely human conscious states and human beliefs and desires, it is very likely that there are things that set them apart from representational contents in non-psychological cases. They have further features relevant to content determination (§8.9), that are missing in non-psychological cases like animal signalling, plant tropisms, and hormonal signalling. My pluralism allows that the right account of content determination applicable to these cases might well be different. The special features of consciousness, or of the practice of offering and assessing reasons for action at the personal level, for example, might well make an important difference to content determination. If so, there will be an account of personal-level content (or more than one) which applies only to psychological cases. However, I would resist the urge to identify these as the real or true accounts of content, since they would not apply to many other, subpersonal, psychological cases, where representational contents do have a clear explanatory role to play. Accounts of content that are plausibly restricted to the psychological are too narrow, and we have seen that accounts of content adequate to capturing the way representational explanations work in subpersonal psychology do indeed extend to some non-psychological cases.

8.6 Development and Content

Various theorists writing about concepts have noted that there is a tight connection between the circumstances in which a concept develops and the object or property it refers to. With representations less sophisticated than concepts a similar connection is often apparent. In previous work on artificial neural networks I explored the way that vehicles of content themselves develop as a result of training, that is as a result of reacting to samples in the environment and being tuned based on feedback about performance (Shea 2007a). This developmental process leads clusters to form in hidden layer state space. Those clusters are vehicles of content. They represent properties of the samples that caused their development.

If content is fixed by synchronic properties of a system, for example its causal sensitivity, then it is far from obvious why there should be any connection between the circumstances in which a representational vehicle develops and what it represents. Fodor was so puzzled by this phenomenon that he gave it a name, the DOORKNOB-*doorknob* problem (Fodor 2008). For Fodor the problem arises because it seems that many concepts are neither innate (i.e. unlearned), nor constructed out of existing concepts. (The problem is discussed in Shea 2016.) The concept DOORKNOB is neither present at birth, nor is it plausibly constructed out of other concepts like ROUND, ATTACHED TO A DOOR, FOR TURNING, etc. Instead, it is acquired by a thinker as a result of experience, a particular kind of experience: interaction with . . . doorknobs. Fodor was puzzled as to why that should be; puzzlement that arises in part, I suggest, from implicitly rejecting the idea that the circumstances in which the concept develops could play a role in fixing its content. After all, for Fodor a knock on the head could fortuitously put a thinker in a new brain state such that they have the concept DOORKNOB.

There are many empirically studied cases where the causes of the development of a new representational resource figure in its content. An example is recognizing new people by their faces. We acquire the ability to recognize an individual by seeing and perhaps interacting with that individual for a short time. The new recognitional capacity that results is probably dependent on a neural representation in the fusiform face area of the brain (Kanwisher 2000, Cohen and Tong 2001). The person who caused the development of this new recognitional capacity, which is mediated by a new vehicle of content, ends up being its referent. This is very similar to my neural network example

(Shea 2007a), where new state space clusters represent properties of the samples that caused their development. Game-theoretic sender–receiver models also show how new representations arise as a result of stabilizing processes (Skyrms 2010); for example, the replicator dynamics can lead senders and receivers to coordinate on a way of categorizing a range of stimuli (O'Connor 2014).

Laurence and Margolis (2002) have an account of the acquisition of natural kind concepts that links their content closely to the circumstances of their development. The child develops a new natural kind concept as a result of seeing a member of the kind:

> She sees a new object that has features that suggest that it is a natural object of some sort... upon encountering the item, the child releases a new mental representation and begins accumulating information about the object and linking this to the representation. [2002, p. 42]

So, an object which falls under a new kind causes the child to acquire a representation which refers to the kind, a representation which is used to store information about the new kind. Laurence and Margolis picture a process in which an existing contentless mental symbol is taken off the shelf and put into the right mind–world relations to constitute it as a concept of the kind. In my neural network case, interaction with samples in the world causes the development of a new vehicle which has the appropriate mind–world connections to have a certain content. Vehicle development and content development occur in tandem, due to the same causal process. In both cases there is a tight connection between the circumstances of development and the content of the new representations that result.

My accounts of content make that conclusion entirely unsurprising. Contents are fixed relative to task functions. Task functions arise as a result of some stabilizing process. Learning is a key case. I argued that outcomes that are the target of stabilizing processes are often stabilized and robustly produced as a result of internal mechanisms, mechanisms that make use of exploitable relations between internal components and the world. One common way that can happen is when the stabilizing process—for example, learning—gives rise to the internal mechanism which is responsible for an outcome being produced robustly and stabilized. Since contents are fixed by reference to outcomes that were stabilized and conditions that obtained which explain why those outcomes were stabilized, it is entirely unsurprising that contents should often concern properties of the objects the system was interacting with during development of a new representation (i.e. during the process of stabilization, that being the process by which content is constituted).

So, a feature which proponents of a synchronic metaphysics of content need to explain, and which poses a puzzle for Fodor, turns out to be readily explicable in the varitel framework. When a new representation develops as a result of interactions between a system and its environment, it will often end up representing the objects and properties causally involved in its development.

8.7 Miscellaneous Qualifications

In this section I go through a miscellaneous series of clarifications and qualifications.

In §2.3, 'Externalist Explanandum, Externalist Explanans', I argued that we should expect contents to be fixed by some complex relational property of representational vehicles. That would make contents suited to explaining how a system achieves distal outcomes in its environment. I have just argued that my accounts fulfil that promise (§8.2). A system will have representational contents based on relational properties that bridge across a variety of different inputs and outputs, and thus are part of world-involving real patterns. Does it follow that contents can only concern distal features of the organism's environment and distal outcomes?

The answer is no. Proximal inputs like sensory properties, and proximal outputs like bodily movements, can be represented. UE information can concern internal states of the system. The point about explanatory purchase requires that the system should have some task functions that concern distal outcomes. It follows that it should have some descriptive representations about distal features of its environment. But that does not imply that every representation must concern something distal. An organism may also represent proximal inputs and outputs, as a means for calculating what is the case and what to do. For example, it may represent possible motor programs and use them in calculating which motor program needs to be executed in order to achieve some directively represented distal outcome in the current circumstances. Or it may keep track of sensory properties as a means for learning how to behave in new circumstances. Whether proximal features are represented in addition to distal ones depends on what is called for by the algorithm the organism is using in order to achieve its distal task functions.[17] Task functions are world-involving, and some representations in a system must be too, but it is not a requirement on being UE information that the correlation should concern a condition distal to the system (nor for UE structural correspondence).

A clear example of that is meta-representation. Some computations call for representations which represent the content of other representations. That arises in the relatively low-level system that does model-free reinforcement learning, since the algorithm involves a stage where the reward expected for an action is compared to the reward actually received and the difference is used to update reward expectations for the future (Shea 2014c). Varitel semantics can readily accommodate, both representations whose content concerns the content of other internal states, and representations that concern non-contentful internal states of the system (e.g. sensory states, bodily states and other internal properties).

A second caveat: I need to qualify the way I have discussed outputs produced by an organism. I have talked as if all outputs are bodily movements or the consequences of

[17] There is a rough parallel here with Burge's view. He argues that the capacity to represent properties like time, for which there is no constancy detection mechanism, is derivative from the capacity to represent properties for which there is a constancy detection mechanism (Burge 2010).

bodily movements. In fact other kinds of output can also qualify; for example, releasing a chemical, producing an electrical discharge, or changing colour. Although movements have taken centre stage in all our discussions, everything I have said should be taken to cover outputs in general (when the other conditions for content are met, e.g. the output is or leads to an outcome which is task function of the organism).

Another significant oversimplification is found in the way I have talked about vehicles of content as constituent parts of an organism or other system. That is, I have pictured them as proper parts, or components of a mechanism. Those are the easiest cases to understand, and this conception covers all of our case studies, however the account is not restricted to such cases. Vehicles need not be proper parts of the system. Syntactic types can depend on properties of the whole system.

I don't know of any actual cases, but for illustration imagine a cell that is simultaneously undergoing three cyclical physiological processes, each something like the Krebs cycle, but involving the whole cell rather than a series of constituents. Take a dynamic systems approach to the cell. Each cyclical process can be occurring in a range of ways, which I'll call states of the cycle. Cycle C can be in various cyclical states C_1, or C_2, and so on; cycle D in states D_1, D_2, etc. Cycle C undergoes changes between cyclical states in a way that is affected by the states of cycles D and E, and vice versa. The whole system exhibits attractors, and perhaps bifurcations, and is affected by states of the environment. Dynamic properties of the whole cell, like being in state C_1 and state D_2, could in principle be vehicle properties, carrying world-involving contents, and interacting in ways that, by obeying generalizations about dynamic interactions amongst the C_i, D_i and E_i, are faithful to their contents. So, vehicles need not be proper parts of the system doing the representing. Furthermore, they need not form a mechanism (assuming not every causal interaction calls for a mechanism).

Next, a brief note on how my approach relates to evolutionary game theoretic models of signalling, communication, and meaning. These models were developed by Brian Skyrms and others (e.g. Skyrms 2010), following David Lewis's decision-theoretic treatment of signalling games (Lewis 1969). For Skyrms, the meaning of a signal in a signalling game is just a matter of the correlational information it carries, in particular how much it changes the probability that world states obtain or actions will be performed.

Shea et al. (2017) argue that these models need to be supplemented with a richer conception of meaning in order to account for phenomena like misrepresentation and deception. These phenomena arise in discussions of the models but are not given a formal treatment. We call this kind of meaning 'functional content', contrasted with the purely informational content put forward by Skyrms. In our treatment, functional content only arises at an equilibrium. It would be possible to apply our definition to non-equilibrium states. Functional content is essentially a matter of how signals are involved in generating rewards in certain world states, given the way receivers act on the signals. Signals are involved with generating rewards whether or not the population is an equilibrium. However, Shea et al. (2017) focus on a kind of content that only arises in an equilibrium state. Varitel semantics is broadly similar. Content depends on task functions. Task functions must be stabilized functions, which means they must have contributed to a stabilizing process like natural selection, learning, or contribution to an organism's persistence (leaving aside task functions based on deliberate design). So, they must, in some broad sense, have been attractors of the dynamic interactions between system and environment. The system need not currently be in an attractor state, but it must have been in one for its states to have content.

Tying functional content to stabilizing processes might seem problematic in the light of recent work showing that, in finite populations, signalling can arise in non-equilibrium states (Wagner 2012, 2015). For example, Wagner (2015) analyses a signalling game where populations converge on an attractor that is not a Nash equilibrium. Senders send signals that are informative about the world state and receivers act on them accordingly. My answer is simple. A stabilized function in my sense need not be a Nash equilibrium. The states Wagner identifies, being attractors to which the model converges, can thereby generate stabilized functions of the sender-receiver system. There may be a legitimate role for defining a notion of functional content for game theoretic models which applies to all states of the game, attractors and transients. However, my framework is motivated by the need to explain successful and unsuccessful behaviour. The parallel in game theoretic models is content that arises from attractor states. So, the restriction to defining functional content only in cases where there is or has been a stabilizing process (attractor) is suited to our purposes.

Moving on, there are some important issues about the nature of content, which should certainly be central to an account of conceptual content, that I have overlooked entirely. One is whether there is a level of content at the neo-Fregean level of sense; for example, a mode of presentation. I have been working only with referential content. Referential content, plus facts about vehicles—for example, that a system has two different vehicles concerned with representing colour—have together been adequate to explain all our target phenomena. Nevertheless, I want to remain neutral on whether a second level of content is justified in our simple cases, or indeed whether it turns out not to be needed even when we come to beliefs, desires, and concepts.

I am also setting aside questions of indexicality. For example, I assumed that particular locations figure in the content of the rat's spatial representations, but have not said whether they are picked out indexically or by non-indexical singular terms (§6.2d). Similarly, in the analogue magnitude case, the monkey's choice between two buckets of objects is correct if it picks the more numerous collection, but I have not said whether the analogue magnitude register for each bucket has the indexical content *that bucket contains n objects*, or whether the collection is picked out non-indexically as in *bucket A contains n objects* (or indeed whether the content is indeterminate between these possibilities). There may be representations which the organism itself reuses in a variety of different contexts, where the right account says that content is a character, that is, a kind of content that combines with a context to deliver a truth condition. I remain neutral about how these issues should be dealt with.

8.8 How to Find Out What Is Represented

My accounts of content say what makes it the case that a simple system has representations with a certain content. It is concerned with the metaphysics of content, not with how we should find out what is represented. Nevertheless, varitel semantics has some straightforward implications for the epistemology of content.

The varitel framework is an elaboration of a procedure that is often used to establish content in cognitive science. Look at patterns of behaviour that are purposeful or adaptive and consider how the organism could perform in those ways by keeping track of aspects of the environment and calculating what to do when. That is, consider what algorithms could be producing the observed behaviour. Then search for evidence about internal workings in order to decide which possible algorithm is actual. Evidence of internal processes may be direct, through imaging, recording, or intervening in the brain; or indirect, through observing patterns of error, interference, and facilitation, like priming effects. When an algorithm which would produce the observed behaviour plausibly maps onto internal workings in the organism, then those elements are good candidates for vehicles of content, and the algorithm tells you what they represent. According to this picture, an early step is to look for robust outcome functions and assess whether they are also stabilized functions, and thus amount to task functions (i.e. outputs that are susceptible to representational explanation).

Considering task functions to be the target of representational explanation is seldom very explicit in cognitive scientific practice, however it is often implicit, regulating which kinds of behaviour are taken to be interesting and in need of representational explanation. More obvious is the search for information. Cognitive neuroscience directs a lot of energy at measuring the correlational information that is carried by different individual neurons, distributed arrays of neurons, and neural areas. My approach implies that not all information is relevant. Only information that is potentially germane to the task should be of interest. That restriction is often implicit in the scientific practice. Single unit recording investigates selectivity in respect of natural features of the distal world like lines, edges, and surfaces. Imaging usually looks for selectivity in respect of tasks or task-related features like faces, locations, object categories, and so on. So, in practice scientists are often in fact only interested in information that is potentially relevant to explaining how the organism behaves.

It is also implicit that information, to be relevant, has to be carried in a way that can be detected by downstream processing. When asking whether some neural area uses a rate code or a phase code, for example, a key consideration is whether the putative code can be read out by downstream processes. Katz et al. (2016) undermined the widely assumed importance of a signal in the lateral intraparietal area (LIP) by showing that knocking it out pharmacologically made no difference to behaviour. Hunt et al. (2012) formulate this requirement explicitly, noting that the information that can be decoded by an observer recording with an imaging technique or electrode can be quite different from 'the functional representations in the network, those used by the brain' (p. 474). This makes explicit a constraint which is usually at work implicitly in cognitive neuroscientific practice.

The varitel framework does recommend some tweaks to current practice. When measuring correlational information, studies focus heavily on input sensitivity. Correlations with actions and outcomes are not completely overlooked, but they deserve greater emphasis, since output correlations almost always play a role in content determination. Furthermore, there could be a more explicit focus on the way behaviours are stabilized so as to underpin stabilized functions. In reward-based learning experiments, input correlations with reward delivered are always considered, output correlations with reward generated less so, although those output correlations are equally relevant. Indeed, it may be possible to generate quantitative measures, not just of correlational information, but of the way representational vehicles are involved with generating variable amounts of reward, along the lines of the reward-involving functional content vector defined by Shea et al. (2017).

Another mainstream way that content is investigated sits very naturally with varitel semantics. Investigators look for illusions and other systematic patterns of error. One might think that an error can only be identified once representational content has been ascertained, but in many cases problems at the level of behaviour, like vacillation, delay, or doing something clearly maladaptive, can be identified before being traced back to errors in what is being represented. My framework shows how we can be more rigorous about success and failure of behaviour. We need to consider the stabilization processes that have been at work; that is, what the organism has evolved and learned to do, and how its behaviour contributes to its persistence. These provide the standards against which success and failure of behaviour and its consequences should be judged. It follows that appeals to ethology and comparative psychology are of more than background interest. They throw light on evolutionary functions, which play a central role in constituting content.

8.9 Differences at the Personal Level

The book has not attempted to tackle personal-level contents. We have focused instead on the hopefully simpler question of how content arises in subpersonal representational systems. In this final section we look briefly at how various features of the personal level might make a difference to content determination.

First off, consciousness. The phenomenal character of conscious mental states may be fixed by intrinsic properties of the subject. More controversially, consciousness may in turn determine the representational content of those states. A naturalistic theory of consciousness seems a distant prospect. If so, we may be a long way from a theory of content for conscious states. On the other hand, there is some hope that the representational content of conscious states may determine their phenomenal character, in which case a theory of content for conscious states will be a route to a theory of consciousness. To follow that route, we need a better understanding of the distinctive functional role of consciousness in order to see which aspects of the way conscious states operate may play a content-determining role. Relevant functional features could include: a global workspace, a drive for consistency between information from different modalities and subsystems, integration of descriptive information with valence and motivation, practical grasp of the enabling conditions for forming a representation reliably, a role in practical reasoning and learning for the future, storage in episodic and semantic memory, and feelings of confidence. Any or all of these features may play a role in content determination for conscious states. None is obviously reducible to the ingredients we have been working with so far.

A second potentially relevant feature is meta-representation or metacognition. On some views a thinker's object-level conscious state of seeing a red rose is simultaneously a meta-level state with a content along the lines of *I am currently seeing a red rose*. There could also be non-conscious mental states that have meta-representation built in. In either case, the fact that object-level and meta-level content are fixed in parallel would form an important part of the theory of content determination.

Thirdly, many theorists of concepts have thought that relations amongst concepts play an important role in determining content. A concept may encode information about how entities and properties are related causally and hierarchically, for example. This gives rise to deductive and inductive entailment relations between concepts. Alternatively, information about relations between categories may be encoded in a thinker's disposition to draw inferences between concepts. Furthermore, some ranges of properties are mutually exclusive, and some objects physically exclude one another in space. The relations of entailment and exclusion instantiated in a network of beliefs, or of concepts, may play a role in content determination in a way that has not been covered by the framework we have been using.

There may be an important difference between implicit and explicit connections between representations.[18] Suppose that when I see *that is a brown furry object of xyz shape/size* I am thereby disposed to think *that is a dog*.[19] This disposition implicitly encodes the information that objects that are brown, furry and of xyz shape/size tend to be dogs. That implicit representation is true. It's being true partly explains why I successfully behave in a dog-appropriate way on this occasion. My DOG concept also figures in some of my explicit representations; that is, beliefs about dogs. I believe that is okay to leave a docile dog alone with a young child. Suppose that belief is false. That has potentially disastrous consequences once I become responsible for young children. But I can modify my explicit belief by reasoning and reflecting on it. Becoming a parent, I start being aware of reports of seemingly docile dogs attacking when left alone with young children. So, I change my belief. Information represented implicitly in my dispositions to apply my DOG concept can also change as a result of experience. I can retrain

[18] The contrast between implicit and explicit representation I am using here is spelt out more carefully in Shea (2015).

[19] *xyz shape/size* is some set of shape and size properties represented in visual experience.

my inferential dispositions. But that is a different process from the way conscious deliberation changes my explicit beliefs. Both kinds of connection may be important in the story about content-determination for concepts. In particular, the special functional role of conscious deliberation in forming and changing explicit beliefs may have a special role in content-determination.

The meaning of beliefs and desires may also depend on interpersonal norms, and/or on the meaning of words, which may in turn depend on social processes (§6.5b). For content determination we would then have to cast the net more widely than a single individual, so as to include culturally based stabilizing processes like the patterns of transmission and use of a word in a social group.

As well as beliefs and desires that figure in episodes of thinking, people also have standing beliefs and desires. I have long believed that Lima is the capital of Peru, even though it is many months since I entertained that thought occurrently (until just now). There may be vehicles of the standing beliefs, stored away in long term semantic memory in the same way as data is stored on a computer disk. Or ascriptionism may be the best account (§1.3); for example, Dennett's intentional stance (Dennett 1981; see also Williams 2016, Williams 2018). Either way, standing beliefs may have observer-relative contents. On ascriptionist views there is no straightforward connection between the contents of standing beliefs and the contents of occurrent beliefs that are tokened in episodes of thinking. The content of standing beliefs could be observer-relative while occurrent states of thinking have non-observer relative contents (perhaps fixed in part by social processes, in the way just suggested). These are important content-relevant features of standing beliefs.

Even if additional functional features of the personal level play a content-determining role, should we nevertheless expect the overall varitel framework to apply, perhaps with an augmented menu of exploitable relations? Or, more minimally, should we expect that the explanatory purchase of representational content will still depend on a convergence between task functions and internal workings which form an algorithm for their achievement? The answer is I don't know. It may do. But the richer features found at the personal level may underpin a different kind of account of content constitution. For example, if consciousness, fixed by intrinsic properties, fixes the content of conscious states, then content determination there works quite differently.

A natural thought here, but one that should be resisted, is the idea that subpersonal contents are picked up and used by personal-level processes. Subpersonal vehicles are constituted as contentful because of their relations: relations to features of the distal environment, to outputs of an algorithm implemented in the organism, and to stabilizing processes that have operated on the organism. Personal-level processes may capitalize on some of those same relations, for example the correlational information carried by a concept may be important to fixing its content. But that is not to make use of the content of a subpersonal representation. The contents of subpersonal representations are not things that are sitting around ready to be used by personal-level processes. It is a mistake to think of the content of a vehicle as a property that is routinely portable, a property that would automatically be carried around if that vehicle is deployed in a different cognitive process. The absence of a straightforward connection here means there is no simple way that personal-level contents are determined by contents of any subpersonal representations they make use of. On the other hand, it frees our theorizing about subpersonal content from the need to play a role in accounting for personal-level content.

So I am remaining open-minded about what kinds of insight, if any, varitel semantics will offer into the nature of personal-level content. Might pluralism, at least, come in handy at the personal level? Not just a pluralism which allows that content at the personal level will be different, but a pluralism that expects different kinds of personal-level state to have their content determined differently? That, too, is open to debate. Close connections—for example, between belief content and the content of conscious states—may make it inappropriate to leave pluralism open about content determination as between occurrent beliefs and conscious states.

While I think it's too soon to venture an opinion about how the personal-level story will go, I would argue that the varitel accounts of content represent a substantial advance. We started the book with the question 'what is a thought process?', and with the worry that the powerful answer offered by the representational theory of mind would be undermined if we were unable to answer the related question 'what is representational content?' Now we have an answer to the content question that works for large swathes of the cognitive sciences. Our optimism that we can answer the same question for representation at the personal level should therefore increase.

And it's not just general optimism that naturalism is taking us in the right direction. The varitel accounts of content give us a staging post, a fixed point from which to build. Intentionality is less mysterious now that we can see how cases of it arise through the coming together of some relatively well-understood natural properties. Psychology and cognitive neuroscience have excavated the computational processes that underlie some quite complex patterns of human behaviour. The foundational worry that those theories are based on a false assumption about meaning can now be assuaged. We can see correlation, correspondence, and function at work in these cases, giving rise to representational content in a completely un-mysterious way. So, we are now in a position to ask what needs to be handled differently to deal with personal-level cases. We have a reasonably detailed understanding of how personal-level representations are different, in ways that are relevant to content, as even the brief discussion above indicates. So, we have a good list of resources to draw on. Rather than being stuck at an impasse, with the lurking suspicion that the question is intractable, or representationalism entirely misconceived, we are now faced with a workable research programme—substantial and challenging, but with a clear sense of how to make progress.

That is a valuable payoff; however, the most important achievement of varitel semantics, if it succeeds, is to elucidate the nature of content in subpersonal cases. Subpersonal representation is a big challenge in its own right. The manifest success of the cognitive sciences has seen representational theories deployed ever more widely. It is now pressing to understand the intentionality at the heart of these theories. I have argued that varitel semantics allows us to understand how those explanatory practices work. Huffing and puffing with information, function and structural correspondence can do the job. We can indeed give a naturalistic account of content in the brain and other subpersonal representational systems.

Paragraph-by-Paragraph Summary

Each entry below summarizes a paragraph of the main text, followed by the page number where that paragraph is found.

PART I

Chapter 1—Introduction

1.1 A Foundational Question

Announcing a mystery: what is thinking, what is a thought process? 3

To Descartes, producing appropriate strings of words was just as mysterious as free will or consciousness. 3

Computers do that. 3–4

Insight: thinking is processing of mental representations. That is RTM. 4

This isn't an agreed answer to our problem because we don't know how mental representations get their meaning. 4

There is good evidence for the claim about processing; but I agree we don't fully understand meaning. 4

Fig. 1.1 Babbage's difference engine. 5

The question is: how do mental states get their aboutness? 5

That is the 'content question'. 6

Mental representation is still a powerful idea; there is no serious reason to doubt their existence. 6

But the lack of an answer does lend support to eliminativism, which would be a mistake—it misses something explanatorily important. 6

Fig. 1.2 Typical explanatory scheme from cognitive neuroscience: computation, implemented in brain areas, to perform a task. 7

We leave aside consciousness (too hard). There is non-conscious information processing to get to grips with. 7–8

An answer will also be useful for content disputes in cognitive science. 8

This chapter is a broad overview of existing treatments of the problem (with little argument). 8

1.2 Homing In on the Problem

Brentano identified the problem of intentionality: how can thoughts be about things in the world? 8–9

Intentionality of words may derive from our thoughts. The same story cannot apply to our thoughts, on pain of regress. 9

Giving the content of a concept in terms of perceptual states calls for underived intentionality there. 9

We need an account of underived content determination, in the metaphysical sense—a meta-semantic theory. 9–10

I am offering an answer to both the content question and the representational status question at the same time. 10

A theory of content should be consistent with and if possible illuminate behavioural explanation: correct representation explains success, misrepresentation failure. 10

Misrepresentation is puzzling: if it makes no difference to processing, how can it make a difference to explanation? Our theory should show how. 10

The theory needs to be applicable to psychological cases and to deliver determinate contents there. 10–11

The determinacy problem has incarnations as the proximal/distal, qua, and disjunction problems. 11

We are aiming for naturalism: an account in non-semantic, non-mental, and non-normative terms. 11–12

1.3 Existing Approaches

This section discusses existing approaches to content and the main challenges they face. 12

We can appeal to correlations and mathematical information theory (as founded by Shannon). 12

Correlational information on its own is not enough: not determinate; the strongest correlation may not be the right one; and it does not rule out disjunctions. It is just one ingredient. 12–13

Inferential role semantics faces problems of holism, and of identifying sets of inferences that are shared and individuative. 13

Structure mirroring structure is another candidate. Too liberal on its own, but another plausible ingredient. 13–14

Davidson: having beliefs and preferences is a matter of being so-interpretable. 14

Dennett: the intentional stance is applicable because of real patterns, and so is realist. 14

I reserve 'realism' for cases where there are real vehicles of content, which is our target, and explains more (see §2.6). 14–15

1.4 Teleosemantics

Teleosemantics is my main precursor. It founds content on etiological functions. Leading theories are also committed to an identifiable representation-consumer playing a content-constituting role. 15

Commitment to consumers defines the 'basic representationalist model': a consumer conditions its behaviour on an intermediate representation. 15–16

Fig. 1.3 The basic representationalist model. 15

Content of representation R = condition presupposed given the behaviour output by a consumer in response to R. 16

Millikan's version: content = success condition = condition which explains why the prompted behaviour led systematically to survival and reproduction. 16

This is illustrated by the honeybee nectar dance. 16–17

Fig. 1.4 Honeybee nectar dance. 17

Millikan's 'least detailed Normal explanation' of the pattern of dances homes in on a particular condition for each dance type, e.g. there being nectar 400 metres from the hive. 18

This excludes various general and background conditions from the content. 18

Teleosemantics is the core of a plausible account of content in animal signalling and some other simple cases. 18

1.5 Challenges to Teleosemantics

This section discusses challenges faced by teleosemantics, so my later accounts can be assessed against them. 18–19

First challenge: leading versions require an identifiable consumer whose outputs can play a content-constituting role. 19

It is especially hard to identify consumers in the case of neural representations. 19

Second challenge: to formulate a notion of etiological function that is suited to playing a content-determining role, general enough to cover the cases, and specific enough to do explanatory work. 19–20

Fig. 1.5 Complex functional interconnections in the brain. 20

Relational evolutionary functions do not deliver very specific functions when the function of the learning mechanism is very general (e.g. classical conditioning). 20–1

Instrumental conditioning can found functions in its own right, irrespective of its evolutionary function, as in Dretske (1988). 21

The challenge is to generalize this, and to show the right way to delimit the kinds of etiological functions that should figure in a theory of content. 21

Third challenge: swampman, which raises the question: why should content depend on history? 21

Behaviour is caused by vehicle properties; therefore we could predict the behaviour of an intrinsic duplicate. 21–2

Millikan argues that inductions are based on historical kinds. But that does not explain why we couldn't do induction on synchronic kinds, properties shared between duplicates. 22

Better move: if the explanandum is *successful* behaviour, that is absent in the case of swampman. 22

That does not rescue views based on evolution by natural selection. As a swamp creature interacts with its environment, the explanandum will come into place. 22–3

Fourth challenge: how does content get explanatory purchase? See next chapter. 23

Chapter 2—Framework

2.1 *Setting Aside Some Harder Cases*

Cognitive science relies on representations well beyond the doxastic and the conscious cases. It would be a considerable achievement to give an account of content for these. 25

I set aside four complicating features of everyday representations: consciousness; relations of justification for the person; being reported to others as reasons; and requiring natural-language-like structure. 26

I use 'subpersonal' simply as shorthand for representations where these four features are not important to content determination. 26

We now have a wealth of data about representation in the brain against which to test our theories. 26–7

> Fig. 2.1 A typical case where neural areas are picked out both anatomically and in terms of what is represented and computed. 27

Neural representations are central cases for us, but the aim is for an account that applies widely in the (subpersonal) cognitive sciences. 27

The overall strategy is to start with the subpersonal, both because this dissolves some of the puzzle of intentionality in its own right, and as a significant step towards a theory of the more complex cases. 27–8

2.2 *What Should Constrain Our Theorizing?*

We should not rely on intuitions, especially in subpersonal cases. 28

A theory of content should show how correct representation explains successful behaviour and misrepresentation explains unsuccessful behaviour. 28

That requires us to get into the details of a wide range of case studies from cognitive science. 28

Our theorizing is constrained by the need to make representational explanation intelligible. 28–9

> Desideratum: an account of content should show why representational properties underpin better explanations of behaviour than would otherwise be available. 29

Content has to be more explanatory than an explanation factorized into a causal chain from input through internal working to output. 29–30

E.g. with a rifle, if the firing pin has content, that explanation marches exactly in step with one in terms of non-representational properties of the causal chain. 30

Demanding that representational explanations be indispensable is too strong. 30–1

Fig. 2.2 The mechanism for firing a rifle. 30

The task is to come up with a definition that picks out properties which are realized in the natural world, and to show why the possession of those properties by internal vehicles affords behavioural explanation. 31

2.3 *Externalist Explanandum, Externalist Explanans*

A commitment to vehicles of content—mental representations as physical particulars—means there will be a full non-semantic causal story. 31–2

For an externalist explanandum—explaining a system's reaction to and distal effects in its environment—externalist contents can provide an explanation that could not be given in terms of intrinsic properties. 32

The interactions (mappings from distal features to distal outcomes) that call for representational explanation are a subset: performing functions. 32–3

Fig. 2.3 'Moniac'—which uses water to compute the way money flows through the UK economy. 33

A function is a mapping. It can be achieved in many ways. 34

An 'algorithm' is the way a system achieves a given function. 34

Transitions called for by an algorithm must map onto internal processing going on in the system, non-semantically described. 34–5

E.g. an algorithm to track surfaces by first tracking colour and motion properties of a portion of surface separately. 35

Which extrinsic properties of vehicles determine content? Exploitable relations they bear to the environment. 35–6

Fig. 2.4 An input–output mapping does not fix what goes on inside the box. 35

The algorithmic explanation calls for a convergence between intrinsic properties that account for the processing with extrinsic properties that account for the content. This is the dual nature of content. 36

2.4 *Representation Without a Homunculus*

It is a mistake to see mental representations as requiring interpretation or an understanding homunculus. 36

Instead, dispositions to produce, process, and act on these vehicles in a certain way, without understanding, is what constitutes them as having content. 36–7

Content arises out of a constellation of interpreter-independent natural properties, so does not depend on the organism being interpretable in a certain way. 37

2.5 *What Vehicle Realism Buys*

There being a fact of the matter about which algorithm a system is following depends on realism about representation, i.e. upon intrinsically individuable vehicles of content. 37–8

Realism predicts that errors produce further errors downstream. 38

Realism explains why representational resources are stable and change piecemeal (when observed). 38

All three patterns of explanation depend on there being non-semantically individuable content-bearers. 38

Vehicles are content-bearers, but vehicle types are not necessarily the content-bearing types; the latter are 'syntactic types'. 39

Vehicles are individual bearers of content picked out in terms of processing-relevant non-semantic intrinsic properties. Syntactic types are types that are processed the same way by the system, and so are guaranteed to have the same content. 39

Syntactic individuation is partly externalist, but still vehicles within a system share processing-relevant properties, which is what underpins the three patterns of explanation above. 39–40

Dynamical properties can be syntactic properties—interacting in ways that are faithful to their contents. 40

Nothing in dynamicism as such counts against dynamical properties being representational properties. 40

A note on metaphysics of mind. A naturalistic theory of content need not be reductive. 40–1

An illuminating sufficient condition for content can admit of exceptions, exceptions that can only be explained at a more fundamental level. 41

Nor need we get a necessary and sufficient condition for content. A series of different sufficient conditions applicable to different cases would be fine. 41

2.6 Pluralism: Varitel Semantics

My framework has two variable elements: functions and exploitable relations. 41

Exploitable relation is a disjunctive category for me, with no unifying cover-all definition. If a condition that covers both correlation and structural correspondence were too liberal, that would rob representation of its distinctive explanatory purchase. 41–2

Functions are a second source of pluralism. Any theory that covers all four types may well over-generate. 42

Since the definition of task function is disjunctive, it is not really a single condition. 42

A final source of pluralism is that content may well be constituted differently at the personal level. 42

I am not here making a positive argument for pluralism, but resisting the call for a cover-all necessary and sufficient condition. 43

'Varitel' semantics marks the variety of exploitable relations, and of functions, hence the varietal set of accounts of content. 43

Distinctive features of the view: pluralism; restriction to the subpersonal; no consumer; tested not by intuition but another desideratum; and my take on realism, exploitable relations, and the dual nature of content. 43

PART II

Chapter 3—Functions for Representation

3.1 Introduction

This chapter focuses on the functions that underpin content, 'task functions'. These are characterized, not by relation to intuition, but by the need to underwrite the explanatory role of representation. 47

Swampman, and the fact that the evolutionary functions of some learning mechanisms are highly unspecific, both motivate a notion of function-for-representation wider than just evolution by natural selection. 47–8

My account combines two elements: producing an outcome in a range of circumstances (robustness) and having a consequence etiology (stabilization). 48

3.2 A Natural Cluster Underpins a Proprietary Explanatory Role

Outcomes being robustly produced and the target of stabilization processes are two features that cluster naturally with a third: having internal representations that produce such outcomes. 48–9

Producing robust outcomes is often non-accidental. It occurs for a reason. 49

Survival of an organism, with its behavioural dispositions, is a fundamental robustness tactic. 49

Learning is nature's other great robustness trick, a route to achieving important outcomes more robustly. 49

An outcome F's having been the target of these stabilizing processes (natural selection, learning and/or contribution to the persistence of an organism) naturally clusters with F's being robustly produced. 49–50

Food retrieval in mountain chickadees is a robustly produced outcome that has been the target of all three stabilizing processes. 50

Stabilization and robustness come together to constitute task functions. 50

How are outcomes produced robustly? 50–1

Often that is due to processing over internal states, internal components that stand in exploitable relations to relevant aspects of the environment (i.e. over representations). This forms the third element of our natural cluster. 51

This cluster supports a host of defeasible inferences. It forms a natural kind. 51–2

Fig. 3.1 Clustering of these three features (schematic). 52

3.3 Robust Outcome Functions

Robust outcome functions, roughly, are outcomes that result from behaviour we are inclined to see as being goal-directed. 52–3

Motor control is a good case where we know about the synchronic and diachronic mechanisms responsible for producing outcomes robustly. Online control tunes the action during execution. 53

Diachronic mechanisms recalibrate reaching dispositions when input or output is altered (e.g. by prism goggles or an artificial force field). 53

Fig. 3.2 Subjects adjust their reaching trajectory during action execution. 53

This case illustrates two key features: the outcome is produced in response to a range of different inputs and the outcome is robust across perturbations during execution (i.e. in external circumstances that obtain). 54

Some propose an additional requirement: that the organism should use different means in different circumstances. 54

Stabilizing processes can result in a cover-all strategy, where an outcome is produced by just one means, and when it is produced is sensitive to relevant external circumstances, so we should not require that robust outcomes must be produced by a repertoire of different means. 54–5

Sensitivity to inputs does not imply that the organism has to be interacting with an object on which its behaviour is targeted (a requirement that is a weakness of cybernetic accounts of goal-directedness). 55

Pursuing a single means whatever the circumstances does not count. 55

Definition: robust outcome function. 55

'Output' covers bodily movements, actions, and their consequences. 55–6

What counts as a different input for clause (i), and not just a different instantiation of the same type of input, is a subtle issue. 56

The idea of 'relevant' external conditions in clause (ii) also needs careful handling. 56

3.4 Stabilized Functions: Three Types

(A) CONSEQUENCE ETIOLOGY IN GENERAL, AND NATURAL SELECTION

Stabilized functions are based on natural selection and/or learning and/or contribution to an organism's persistence. 56

'Consequence etiology' is where an output F is generated because of its consequences: F occurs because it is produced by a system S, and S is present because it produces or has produced F. 56–7

Fig. 3.3 General treatment of consequence etiology. 57

This draws the category of function too broadly for our purposes. 57

I draw it disjunctively. Evolution by natural selection is the first case. 57

(B) PERSISTENCE OF ORGANISMS

A ubiquitous enabling condition for producing an outcome F robustly is the persistence of an organism disposed to produce F. 57–8

Several accounts of biological function base it in contribution to a system's persistence. 58

To define the class of systems, contributions to whose persistence should count, I help myself to the category *organism*. 58

Bacterial chemotaxis exemplifies contribution to the persistence of an organism. 58–9

An outcome F which is like that occurs partly because of the consequences which flowed from the individual organism producing F in the past. 59

The effect is not specific to F (unlike natural selection or learning), since it preserves S with all its behavioural dispositions. Persistence is an indirect route, when combined with learning, towards making F more robust. 59

(c) Learning with Feedback

Learning need not be driven by outcomes that contribute to persistence of the organism. 59

With feedback-based learning, the consequences of producing F (e.g. of pressing a key) account for the disposition to produce F (in certain circumstances), and to do so robustly. 59–60

Fig. 3.4 Subjects learn from feedback to make reaching movements that maximize reward. 60

Learning is a form of diachronic flexibility in the service of being able to achieve important outcomes robustly. 60

Learnt behaviours have derived evolutionary functions, derived from the function of the learning mechanism, e.g. to track people by their faces. 60–1

Derived evolutionary functions can be quite unspecific, e.g. when learnt by classical conditioning. 61

A learning process can explain stabilization without explaining why certain outcomes (e.g. getting money) reinforce/modulate dispositions to behaviour. 61

This encompasses one-shot learning, negative reinforcement, and cases where achieving outcomes close to O increases the organism's tendency to achieve O. 61–2

Behaviour learnt by imitation may involve feedback, or may have a stabilized function as a result of cultural transmission and selection. 62

In sum: feedback-based learning is a third source of stabilized functions. 62

(d) A 'Very Modern History' Theory of Function

Stabilizing processes are not like kinematic equilibria: forces holding a pattern in place at a time. 62

So, it is tempting to treat them counterfactually or in terms of what is likely to be stabilizing. 62

Whether an outcome will contribute to persistence (or learning or natural selection) is unsuitably open-ended. 62–3

Forward-directed functions are also poorly suited to being part of an explanation of why outcomes are produced; historically based functions can. 63

So, I define stabilized function in terms of an actual causal history of natural selection, learning or contribution to an organism's persistence: a 'very modern history' theory of function. 63

Definition: stabilized function. 64

Evolution covers cases of cultural transmission and selection. 64

Outputs on the way to, or downstream of, some output F that is stabilized do not thereby also count as stabilized functions. Outcomes that have contributed to stabilization unsystematically also don't count. 64

3.5 *Task Functions*

The functions-for-representation that will figure in our theory are task functions. A task function is a robust outcome function that is also a stabilized function or the product of intentional design. 64

Functions based on intentional design do not meet our criteria for naturalism, so while noting that they share features with the others, I mostly set them aside. 65

I also set aside representational contents that are the direct result of design, as when a designer intends that x should represent y. 65

Definition: task function. 65

This is not offered as a definition of biological function, and I don't follow previous theorists in aiming to reduce the supposed normativity of content to the supposed normativity of biological function. 65

Task function works in the subpersonal cases we will examine. Other notions of function (or none) may be needed elsewhere. 65–6

Task functions vary in ways that affect the explanatory bite of the contents they generate. For example, robustness and stabilization come in degrees. 66

They also vary in how many of the alternative bases of stabilized function are present: all three in paradigm cases, but they need not line up and may underpin different (and contradictory) contents. 66

In the balance between generality and informativeness, our cluster is encountered often in nature while still having the benefit of underpinning a rich set of inductions. 66

3.6 *How Task Functions Get Explanatory Purchase*

(A) ILLUSTRATED WITH A TOY SYSTEM

This section looks at a toy system based on the comparator circuit for motor control of reaching. 67

Our system will move along a line, from a range of initial positions, to a set position *T*, where it stops. 67–8

Fig. 3.5 The toy system. 67

We can explain how the system achieves that outcome by reference to the processing of internal components that bear certain exploitable relations to features of its environment. 68

Reaching *T* is a robust outcome function. 68

If it recharges at *T*, then reaching *T* also becomes a stabilized function; also if it learns, so that recharging cements an initially randomly set behavioural disposition. Reaching *T* is then a task function. 68

An interaction between four internal components, given how they correlate with things in the environment, explains how the system achieves this task function. 68–9

Successfully reaching *T* is explained in terms of components correlating as they did during stabilization, i.e. representing correctly; failure to reach *T* is explained in terms of misrepresenting. 69

(B) SWAMP SYSTEMS

What if our toy system were assembled by chance? It would have a robust outcome function, e.g. reaching the location *T*, but none of the possible robust outcomes seems to count as a success. 69

Suppose there happens to be a power source at *T* and the system has been moving around for a while, during which time it has recharged at *T*. Reaching *T* again would then count as a success. 69

Going beyond that intuition, we can see that the success/failure distinction arises because of our cluster. Then the successes are outcomes for which it is explicable both why and how they are produced. 69–70

The same goes for organisms: a robust outcome function of a swamp monkey would not count as success or failure until the individual's interactions with the world had contributed to persistence or learning. 70

These thought experiments illustrate how task functions, and the success/failure distinction, can arise independently of evolutionary history. 70

Learning-based task functions are the same, e.g. clapping because it makes a parent smile. 71

Task functions still depend on history, but unlike standard teleosemantics, not on deep evolutionary history. 71

The robust outcome aspect of task functions means that there are real patterns in system–world interactions that generalize across distinct proximal properties. That contributes to the proprietary explanatory purchase of representational content, cp. §8.2b. 71

Unlike in the case of the rifle firing pin (§2.2), robust outcome functions 'bridge' to common outcomes across a range of proximal conditions. 71

Standard teleosemantic accounts do not include a robust outcome requirement, although their motivating examples would support it. 71–2

3.7 *Rival Accounts*

Griffiths argues that function should not be analysed in terms of contribution to persistence, because natural selection is aimed at reproduction and that can work against persistence of the individual.72

Our account still applies in such cases. Task functions will be based on evolution rather than persistence. 72

Griffiths offers a rival forward-looking account. 72–3

The objections above to forward-looking accounts (§3.4d) apply equally here: they are too open-ended, and are unsuited to figure in a causal explanation of behaviour. 73

A consequence of my account is that representations based on an evolutionary history of stabilization (reproductive fitness) can conflict with those based on persistence of the individual (survival). Task functions can also arise based on contribution to persistence without any evolutionary benefit. 73–4

3.8 Conclusion

Three features cluster together for a natural reason: robustness, stabilization, and processing over internal components bearing exploitable relations. This constitutes some behavioural outcomes as successes and allows us to explain both how and why they are produced. It is this cluster that gives rise to the kind of representational content found in our case studies. It allows us to see why content has a distinctive explanatory role, thus satisfying our desideratum (§2.2). 74

Chapter 4—Correlational Information

4.1 Introduction

(A) EXPLOITABLE CORRELATIONAL INFORMATION

The next two chapters turn to the internal algorithm; in this chapter, where correlation is the relevant exploitable relation. 75–6

The account does not call for dedicated representation consumers that play a content-constituting role. 76

Definition: correlational information. 76

Correlations are useful. Behaviour can be conditioned on a correlate. 76

The exploitable correlations—ones that can be relied on, and where things going well is explicable rather than accidental—are nomologically underpinned by a single reason. 76–7

Definition: exploitable correlational information. 77

The regions in which a correlation subsists can be very local. 77–8

Exploitable correlational information may be carried by a range of states about a range of states. 78

Definition: exploitable correlational information carried by a range of states. 78

Cases of exploiting a correlation: animal signalling; equilibria in Skyrms-style signalling games. 79

Exploitable correlational information is liberal: it exists with respect to many different regions, with different strengths. 79

The relevant correlation strength is that within the region encountered by organism/system (individual or type). 80

(B) TOY EXAMPLE

We will look at a toy example to see the way correlations are exploited. 80

Table 4.1 Useful correlational information carried by the four internal components. 81

Fig. 4.1 Toy system discussed in the text. 81

Given the correlational information carried by the internal components, the way they are processed constitutes a simple algorithm for moving to a power source. They meet our desideratum for the explanatory role of content. 81

The components carry other information, e.g. about sensory stimulation or rotation speed of the wheels, but that information cannot explain task performance so directly. 82

Correlation of internal components with light intensity would also offer a less direct explanation. 82

Some content indeterminacy remains, e.g. as between distance to a power source and distance to something worth reaching. 82

4.2 *Unmediated Explanatory Information*

(A) EXPLAINING TASK FUNCTIONS

Our motivation for representationalism implies that the content-constituting correlations are those which explain S's performance of task functions. 83

We are not asking which content assignment gives the best representational explanation of behaviour. We are asking which correlations offer the best causal explanation of robustness and stabilization. 83

To explain S's performance of task functions is to explain how outcomes were stabilized and robustly produced. 83–4

Definition: explanandum: S's performance of task functions. 84

Definition: unmediated explanatory information (UE information). 84

A correlation with C plays an unmediated role in the explanation if its role does not depend on C correlating with some further condition C'. 84

UE information constitutes content. 84

Condition: for content based on correlational information. 85

The theory closely parallels the use of model-based fMRI to find representations in the brain. 85

To choose between candidate algorithms that would account for the observed behaviour, we examine which algorithm best accounts for trial-by-trial variations in neural activity. 85

As well as keying into external conditions, explanatory correlational information also has to fit with internal processing: with (locally caused) transitions between the information-carrying vehicles. 85–6

Algorithms usually call for processing steps in which different vehicles have different contents. 'Unmediated' does not count against the computation being mediated by a series of steps. 86

UE information covers output correlations too. These give rise to directive contents. But there has to be some descriptive content in the system somewhere. 86

There are some internal states whose tokening correlates with achieving an output that is itself a task function, e.g. getting sugar. When it forms part of a wider story, that will count as UE information. 86–7

My account is in the spirit of Dretske (1986, 1988), where correlational information is a structuring cause of behaviour. 87

But my account is more general: it is not restricted to instrumental conditioning; correlations need not pre-exist; and it applies when several vehicles need to interact to produce the stabilized behaviour. 87–8

The latter is important because real cases have many representational vehicles that interact in complex ways. 88

(B) RELIANCE ON EXPLANATION

As in other sciences, I am assuming a realist account of causal-explanatory relations. 88

So, it is not an interest-relative matter whether UE information exists. It may be interest-relative whether we go in for representational explanation (i.e. appeal to the property we have theorized). 88

If causal-explanatory relations are interest-relative throughout the sciences, then they would be in my theory too. 88–9

We do need to ask whether the property picked out by the definition of UE information is any use. It is, because it is in fact the property that figures in many representational explanations of behaviour. 89

(C) EVIDENTIAL TEST

UE information makes performing task functions more likely. That gives us a test for UE information. 89

Evidential test for UE information. 89

In the toy system, strengthening the correlation of **r** with location directly affects the chance of reaching *T*. 89

Strengthening the correlation with light intensity would have a less strong effect on achieving that outcome. 89–90

The test also applies to correlations at output (e.g. noise in the motor system). 90

It is only an evidential test, not always satisfied, and not necessarily giving the right result. 90

The test is restricted to correlations with conditions involving natural properties, since these are the best candidates to figure in causal explanations of a system's performance of task functions. 90

To apply the test, keep fixed how the vehicles interact, and consider what would happen if the world was in a particular state more or less often. 90–1

The test can be applied to the system in the past (during stabilization) or, less ideally, applied to current circumstances. 91

4.3 *Feedforward Hierarchical Processing*

Here we look at a simple case: the ALCOVE feedforward neural network. 91

Its task function is to put objects of category A into box A. 91

Nodes at the hidden layer ('exemplar nodes') correlate with individual objects and carry information about many other things (i)–(vi). 91–2

Fig. 4.2 The ALCOVE network. 92

At the hidden layer, the unmediated explanatory information is about exemplar identity. 92

According to the evidential test, the UE information at output concerns object categories; at input, perceptual features. 93

At the hidden layer, exemplar and category are tied on the evidential test, but exemplar content gives us a better understanding of how the system achieves its task functions. 93

UE information chooses between coextensive correlations and tends to home in on distal properties, e.g. in another model JIM, the geons are about objects rather than sensory features. 93

A further development of ALCOVE has reciprocal connections. We turn to feedback in §4.8. 93

This account of content makes good on my earlier claim that there is no need for a representation consumer that plays a content-constituting role. 93

4.4 *Taxonomy of Cases*

Neural processing takes place in complex ways. 94

Fig. 4.3 Diagram of the primate visual system. 94

Fig. 4.4 The four kinds of case exemplified in §§ 4.5, 4.6, 4.7, and 4.8 respectively. 95

I will pick out four kinds of cases and look at an example of each. 95

It is very hard to identify a single consumer system when there are feedback loops and no simple hierarchy. 95

In Case 1, one vehicle is used by two different output subsystems. In Case 2, two different representations are used by a single subsystem. 95–6

In Case 3, information is processed via two routes, one direct and one indirect. Case 4 features feedback and cycles. 96

The case studies below consider each in isolation to show that none presents an obstacle to the UE approach. 96

4.5 *One Vehicle for Two Purposes*

It is common to find one animal signal used by different receivers for different information it carries (e.g. mates and predators of a firefly). 96

A cooperative example of that is a chicken signalling to conspecifics and to a predator. 96

Corollary discharge tells the motor system to act and tells perceptual systems that the organism is acting. 96–7

C. elegans contains a simple example of this. 97

Arguably the contents are of different kinds here, descriptive and directive (see Chapter 7). 97

In our case studies, where there is reuse it turns out that the same content is being used by different subsystems. 97

4.6 Representations Processed Differently in Different Contexts

(A) ANALOGUE MAGNITUDE REPRESENTATIONS

The analogue magnitude system looks at first as if it represents different contents in different contexts, but actually there is probably a common representation of numerosity. 97–8

It acts as an (imperfect) correlate of the numerosity of many kinds of array, with discrimination displaying the characteristic Weber's law set-size signature. 98

There is evidence for a common numerosity register—comparisons across modalities, interference, common neural basis—while also registering whether the items are objects, tones, flashes, etc. 98

Consider a stylized system with one register R for numerosity and another R' for stimulus type. 98–9

Fig. 4.5 Case 2. 99

Rather than separate domain-specific contents for different downstream uses, the UE approach implies that R is a common register for numerosity *tout court*. 99

These considerations mean that UE information will often 'triangulate' on a common content. Being decoupled from specific outputs, perceptual representations are pushed towards having purely descriptive content. 99–100

The systematic relationship between activation and numerosity means that the system can represent numerosities beyond those encountered during stabilization. 100

(B) PFC REPRESENTATIONS OF CHOICE INFLUENCED BY COLOUR AND MOTION

We turn to the prefrontal cortex for a case where the two different kinds of information are carried in the same register and used in different contexts. 100

Subjects see an array of random dots in varying proportions of two different colours and with varying motion directions. Their task is either to judge preponderant colour or preponderant motion direction. 100

Fig. 4.6 Behavioural task in Mante et al. (2013). 101

Neural evidence accumulates for both colour and motion, in a distributed neural pattern. 101

The context affects whether the evidence that accumulates to drive choice (movement in the choice direction) is colour or motion. 101–2

Fig. 4.7 Schematic representation of the processing in Mante et al. (2013). 102

Simplify: representational processing of two vehicles at input, one at output (programming saccades). 102

Task function: to get juice; relevant condition is colour in some trials, motion in others. 102

List of correlations with distal features that explain performance of the task function, hence constitute content. 102

Correlations with properties of sensory input would be less explanatory. 102–3

Lumping all the inputs together in a single space is less explanatory of how the system computes. 103

Some indeterminacy remains. That is appropriate to the case. 103

4.7 *One Representation Processed via Two Routes*

Case 3 is where a single representation is processed by two routes, one direct and one indirect. 103

We look at an example of that in the visual system, from van Essen and Gallant (1994). 103–4

Fig. 4.8 Case 3. 103

Fig. 4.9 A detail from van Essen and Gallant (1994). 104

We focus on the interaction between V2 Thin stripe, V2 Thick stripe, and MT, with a simplified computational interpretation. 104

MT detects plaid motion: motion of overlapping surfaces moving in different directions. 104

There are two routes from V2 Thin stripe to MT. 104–5

We consider a simplified system with the task function of intercepting moving objects. 105

MT correlates with direction of motion of encountered objects, V2 Thick stripe with local motion direction, and V2 Thin stripe with a chromatic invariant of surfaces. 105

Different components are doing different jobs. 105

A consumer-based approach could bundle these together into a single intermediate representation, but that would not account for how the system manages to compute motion direction. 105–6

In short, the UE information approach is appropriate for cases of one vehicle processed via two routes. 106

4.8 *Feedback and Cycles*

Bogacz (2015) describes a probabilistic calculation for deciding between options on the basis of sensory evidence. The model involves cyclical information processing en route to action selection. 106

The circuit calculates probabilities that various possible actions will be rewarded. When one of these crosses a threshold the system outputs the corresponding action. 106

To see how the UE approach can apply to probabilistic representations, first notice that fine-grained information about the probabilities of a range of world states can be useful information to have. 106–7

Fig. 4.10 The neural computation proposed by Bogacz (2015). 107

The joint probability distribution of a range of putative representations X and a range of world states Y, if based on a univocal reason, is a *fine-grained exploitable correlation carried by X about Y*. 107

Correlations like that, which figure in explaining performance of task functions, qualify as UE information, hence fix content. 108

Conditions Y about which X carries more mutual information are, *ceteris paribus*, better candidates to be UE information carried by X. 108

We need to replace correctness with a graded notion of accuracy. The Kullback-Leibler divergence of the true distribution from the represented distribution is one appropriate measure. 108–9

In Bogacz (2015), the system takes a sensory input and uses it to calculate posterior probabilities that available actions will be rewarded. If none crosses a threshold, they act as priors that are updated with the next sensory input. 109

The system has been trained by feedback to produce the action that is mostly likely to be rewarded in the current context. 109

Bogacz's probabilistic computational model captures the UE information used by the system to perform the task. 109–10

Fig. 4.11 Case 4. 109

This case shows that the varitel framework can discern content processed in feedback loops. 110

4.9 Conclusion

For a relevant input–output mapping (and Chapter 3 told us that task functions are the relevant ones), content is fixed by exploitable relations carried by components which make internal processing an implementation of an algorithm to perform the input–output mapping. 110

For correlations, the content-constituting ones are those which unmediatedly explain, through implementing an algorithm, how a system performs its task functions. That works in a series of case studies. It does not call for a content-constituting consumer system. 110

Chapter 5—Structural Correspondence

5.1 Introduction

This chapter: structural correspondence is an exploitable relation and can be content-constituting. 111

A correspondence: maps entities from domain 1 to entities in domain 2; and there is a relation between the entities in domain 1 that is mirrored by a relation between the entities in domain 2 to which they map. 111–12

The thin notion of relation makes the existence of a structural correspondence between relations too undemanding. Here I develop principled restrictions to candidate relations on both sides of the correspondence. 112

For every relation on entities in the world, and any way of mapping a set of vehicles onto them, there is a corresponding relation on the vehicles. 112

In general, a structural correspondence of this kind is not something that will help a system perform task functions. So, it cannot be content-constituting while meeting our desideratum. 112

Fig. 5.1 For any relation on entities in the world, there is *a* corresponding relation on vehicles. 113

The cognitive map in the rat hippocampus exemplifies a more substantive kind of correspondence, an *exploitable structural correspondence*. These can be exploited, thus constituting content, or go unexploited. 113

5.2 *The Cognitive Map in the Rat Hippocampus*

Rat place cells fire when the rat is at a specific location. 113–14

Fig. 5.2 Place cells in the rat hippocampus are tuned to specific locations. 114

This is useful information to have, e.g. to learn what to do in different locations, but that would not be to make use of a relation between place cells. 114–15

Taken on its own, place-specific firing does not show that any relation on the place cells is being exploited. 115

Cells for nearby locations tend to activate one another when active offline, showing replay or preplay of routes through space. 115

Co-activation is used to compare different routes to a rewarded location and pick the shorter one. 115–16

This is a matter of achieving task functions robustly, stabilized by interaction with the environment, part of the explanation for which is the structural correspondence between co-activation and space. 116

So, this is a case where use of a structural correspondence to perform task functions is the basis of representational content. 116

5.3 *Preliminary Definitions*

This section defines 'structural correspondence' and 'structural representation', and says what it is for a structural correspondence to be content-constituting. 116

Symbols: the structural correspondence obtains between relation V on vehicles v_i and relation H on worldly entities x_i. 116–17

Fig. 5.1 shows an isomorphism, but I define structural correspondence in terms of the slightly looser notion of homomorphism, which allows for two different representations with the same content. 117

Definition: structural correspondence. 117

This does not imply that the parts have to be representations. For simplicity I use the standard definition of structural representation, which assumes that they are, but my approach could still apply if they're not. 117–18

What it takes to be a structural representation is that a relation on representations represents a relation on the entities represented. 118

Definition: structural representation. 118

We are interested in cases where a relation on vehicles represents a relation in the world *because* the relation on vehicles bears a structural correspondence to the world. The definition of structural representation does not entail that. 118

Definition: structural correspondence as content-constituting. 118

For a system to make use of a structural correspondence, the relation V between vehicles has to make a difference to downstream processing. 119

For contrast, consider vervet alarm calls and, somewhat arbitrarily, the relation H, *higher than*, between predators (i.e. how high off the ground the predator usually is). 119

A relation on the alarm calls corresponds to H, but vervets are not sensitive to that relation (nor to any relation between alarm calls). The structural correspondence exists, but is not content-constituting. 119

The requirement that a structural correspondence be used, therefore usable, cuts down very considerably on the problematic liberality of structural correspondence. 119

5.4 Content-Constituting Structural Correspondence

(A) EXPLOITABLE STRUCTURAL CORRESPONDENCE

This section sets out the substantive notion of structural correspondence we need. 120

In the rat case, co-activation made a difference to processing, and its correspondence to distance was being used. 120

Having a relation that processing is sensitive to correspond to a task-relevant relation in the world is a major achievement. 120

Definition: exploitable structural correspondence. 120

Downstream processing is sensitive to co-activation, but not to the colour of the cell bodies, nor to where they are located within a hippocampal layer. 120–1

The place cells on their own are useful because they enable acquisition of a co-activation structure, but I reserve 'exploitable structural correspondence' for when the relation on vehicles is already in place. 121

NB: the exploitable relation is not the co-activation relation. It's the overall structural correspondence. 121

Neural processing is sensitive to relations between firing rates, and temporal relations between spikes; possibly also to phase offset. 121

Plasticity can alter downstream processing so that a merely potentially exploitable structural correspondence turns into an exploitable one, i.e. one where processing is systematically sensitive to the relation between vehicles. 121–2

The relation on vehicles should make a systematic difference to downstream processing—which can be spelt out. 122

The worldly relation has to be significant for the system, given its task functions. That will usually exclude gruesome and disjunctive properties. 122

Notice that there are different restrictions on the two sides of the correspondence. 122–3

A structural correspondence is instantiated when an instance of the relation V is instantiated between two vehicles, together with an instance of relation H being instantiated between the two worldly entities to which they correspond. 123

We have identified the useful structural correspondences. This captures the sense in which the Survey of India was useful. 123

(B) UNMEDIATED EXPLANATORY STRUCTURAL CORRESPONDENCE

For an exploitable structural correspondence to be exploited is for it to figure in a causal explanation of a system's performance of task functions. 123

Definition:unmediatedexplanatorystructuralcorrespondence(UEstructuralcorrespondence). 123

For the rat, getting to a particular location again is a task function, performed by using the structural correspondence between co-activation and space. 124

This gives rise to content based on UE information and UE structural correspondence. 124

Definition: condition for content based on structural correspondence. 124

So structural correspondence, of an appropriate kind, is part of what gives representations their content. 124

The definition is neutral between descriptive and directive content—see Chapter 7. 124

Exploitable structural correspondence is not (circularly) defined in terms of being exploitable. 125

A UE structural correspondence can determine content about entities x_n and a relation H on them all at once. 125

Fig. 5.3 Points on a simple map pick out locations, and do so in virtue of their spatial relations to other things on the map. 125

A new exploitable structural correspondence can be created by creating new relations amongst existing vehicles, e.g. learning the (arbitrary) count words by rote. 126

Putting together a series of vehicles so that the sequence becomes automatized is a common way to create new structure. 126

In short, new exploitable structural correspondences can be created by constructing new relations on vehicles or by making downstream processing newly sensitive to an existing relation on vehicles. 126

5.5 *Unexploited Structural Correspondence*

In some cases an obvious structural correspondence is not being exploited and is not a basis for content. 126–7

Many deliberately designed representations are set up so as to make an obvious relation usable, e.g. spatial relations. 127

Colour is another easily used relation on representational vehicles. 127

Many cognitive science cases are like the honeybee nectar dance in that a structural correspondence which obviously exists is not being exploited. 127

The bee dance is, however, an 'organized sign system' (Godfrey-Smith 2017). 127–8

Organization in this sense is an important feature. It allows a compact mechanism to extend to a range of cases (cp. §4.1a), and to extend to new cases. It also makes the system error-tolerant. 128

Organization is different from structural representation. 128

Cummins's ingenious driverless car example is a case of structural representation. The structure is the relation between subsequent pin positions on the guide-card. 128

Fig. 5.4 Cummins's case of a driverless car guided by a slot in a card. 129

The distance moved by the pin correlates with how far the car has moved on the ground. The car uses spatial relations between positions on the card to program appropriate actions. 129

Fig. 5.5 One step of the computation being performed in Cummins's driverless car case. 130

This is a case of UE structural correspondence: relations between pin positions on the card represent relations between locations of the car on the ground. 130

Gallistel's concept of a functioning isomorphism inspired, and is much like, my notion of UE structural correspondence, but is more permissive in one important respect. 130–1

Gallistel allows that 'indirect' isomorphisms, 'created only by way of an interpretative code', are a sufficient basis for content. 131

That is too liberal. The interpretative code could work on each representation piecemeal. Then relations between representations would lose any significance for content. 131

I agree that which relations count depends on downstream processing, but this must be a matter of sensitivity in a systematic way to some interpreter-independent relation on the representations. 131

So, I allow some but not all of Gallistel's indirect isomorphisms. 131–2

5.6 Two More Cases of UE Structural Correspondence

(A) SIMILARITY STRUCTURE

We examine two more case studies in which a structural correspondence is exploited and is thereby constitutive of content. The first is exploitation of similarity structure. 132

Similarity in neural activation patterns can be measured by distance in activation state space. 132–3

Fig. 5.6 Illustration of neural similarity space. 132

A neural similarity structure, e.g. as observed in BOLD repetition suppression effects, is relevant if subjects' task is to judge the similarity of objects presented to them. 133

When similarity in neural activation space is being used for the way it corresponds to some objective dimension(s) of similarity between objects, this is a case of UE structural correspondence. 133–4

Note: this is not based on a claim about subjectively experienced similarity. 134

(B) CAUSAL STRUCTURE

The second case involves representation of causal structure, which has been so significant for human evolution. 134

Action selection can be based simply on whether an action led to reward in the past (model-free) or on an understanding of the causal connections between actions and their consequences (model-based). 134–5

The two-step task tests for model-based choice, hence causal understanding, but does not require structural representation. 135

Huys et al. (2012, 2015) tested causal planning using a more complex multi-step task. 135

This causal planning capacity may be an elaboration of the capacity to represent the sequential order of events. When the sequential structure mirrors causal structure, that correspondence is exploitable to do causal reasoning. 135–6

Fig. 5.7 The structure of the task studied by Huys et al. (2012, 2015). 136

If subjects doing causal planning rely on sequential order between brain states corresponding to causal accessibility between world states, then this is a case of UE structural correspondence. 136–7

5.7 *Some Further Issues*

(A) EXPLOITING STRUCTURAL CORRESPONDENCE CANNOT BE ASSIMILATED TO EXPLOITING CORRELATIONS

Often the representing relation will also correlate with what it represents. Is correlation doing all the content-constituting work? 137

Using a relation between vehicles ≠ using it because it corresponds to a relation between the entities represented by those vehicles. Example: the difference between the firing rates of neurons representing the gaze direction of each eye correlates (inversely) with the distance of the object of focal attention. 137

Fig. 5.8 The difference in firing rate between gaze-direction neurons for each eye correlates inversely with the distance to the attended object. 138

But this is being used for its correlation with object distance, not because it corresponds to a relation between the things represented by the two gaze-direction neurons. 138

A second objection is that the cases of computations involving UE information in the last chapter already depend on structural correspondence: that the functional transitions in the computation correspond to the structure of the world. But that is not a case of structural representation at all. 138–9

UE structural correspondence is a special kind of case, with two consequences:

(i) add a new vehicle to the structure and it acquires content irrespective of any correlations;
(ii) the relation can be used to compute across a range of vehicles in a systematic way. 139

An entirely accidentally corresponding structure could be used for its correspondence to the world. 139–40

In sum, UE structural correspondence is a separate basis of content from UE information. 140

(B) APPROXIMATE INSTANTIATION

Correspondences that are only approximately instantiated can still explain performance of task functions. 140

A structural correspondence I is *approximately instantiated* when the actual relation between the objects represented is approximately equal to their relation under I. 140

Since we have put strong constraints on the objects and relation that can figure in I, it is very unlikely that I is ever exactly instantiated. 140

Allowing approximate instantiation opens up a wide class of candidate exploitable structural correspondences. We need to ask how exactly/approximately each was instantiated. 140–1

The degree of approximateness should match the degree to which behaviour contributed to stabilization. 141

This does not trump considerations about which objects and properties are explanatory of task performance. 141–2

Representational redundancy, where two vehicles map to the same object, is possible in a candidate structural correspondence, but this will reduce the accuracy with which the relation between objects in the world is represented. 142

(C) EVIDENTIAL TEST FOR UE STRUCTURAL CORRESPONDENCE

With this notion of approximate instantiation we can formulate an evidential test for content (cp. §4.2): the correspondence whose accuracy of instantiation is most directly connected to the likelihood of achieving task functions is a good candidate to be the content. 142

Evidential test for UE structural correspondence. 142

The test is epistemically helpful in dealing with indeterminacy. 143

The test does not imply that the more accurate correspondence is always the better candidate. 143

Applied to the similarity space on bird images in Constantinescu et al. (2016), the evidential test delivers the content assignment argued for above. 143

5.8 *Conclusion*

The liberality of isomorphism is a problem because it would rob representation of its explanatory power. From our perspective, liberality is a symptom of a deeper problem: most correspondences are not usable, let alone used. This chapter delimited a restricted class of exploitable structural correspondences. We saw that these are a plausible basis for content determination. 143–4

PART III

Chapter 6—Standard Objections

6.1 *Introduction*

The aim of this chapter is to make more explicit the way varitel semantics deals with standard philosophical challenges faced by theories of content. 147–8

6.2 *Indeterminacy*

(A) ASPECTS OF THE PROBLEM

The plan is to look at determinacy problems in the standard case (the frog prey-capture system), the analogue magnitude system and, in subsection (d), cognitive maps in the rat. 148

We'll work with a simplified version of the frog's tongue-dart prey-capture system. 148

Consider representation R, in the retinal ganglion, which responds to flies at location (x,y,z) and causes a tongue dart to that location. 148

Distality problem: does content concern the fly and the action of catching it, or more proximal stimulation and bodily movement? 148

Specificity problem: which of a nested set of co-instantiated properties is represented? 149

Disjunction problem: why is content not a disjunction of relevant conditions? In my terminology, all three are aspects of the determinacy problem. 149

Answers are tested, not against intuitions about the case, but by whether they deliver the right amount of determinacy for explaining content-based explanation of behaviour. 149

(B) DETERMINACY OF TASK FUNCTIONS

First move: task functions contribute some determinacy. 150

Task function and a (simple) algorithm are in place in the case of the frog. 150

Its task function is to catch flies, not little black moving things—based on which properties are directly responsible for survival and reproduction (selection for). 150

This also counts against background conditions (e.g. strength of gravity) and fine-grained qualifications (e.g. not poisonous) figuring in the content. 150–1

The task function is however indeterminate as between fly (biological taxon), flying nutritious object (ecological category) and object worth eating. 151

In the analogue magnitude system, its (learning-based) task functions home in on *numerosity*. 151

(C) CORRELATIONS THAT PLAY AN UNMEDIATED ROLE IN EXPLAINING TASK FUNCTIONS

The requirement for unmediated explanation of task functions also contributes some determinacy. As a result, distal conditions are often better candidates. 151–2

In the analogue magnitude system, correlation with numerosity explains directly. Correlation with sensory properties figures only in a mediated explanation. 152

Asking how a collection of correlations explains an ensemble of task functions also helps, e.g. it homes in on flies rather than food in general. 152

Varitel semantics cannot appeal, for determinacy, to facts about what the organism can discriminate; nor to which correlation is strongest. 152

Convergence between correlational information and task function provides some constraint: exploitable correlational information must be nomologically underpinned, for a univocal reason. That favours a non-disjunctive category like *fly* over a disjunctive category like *species 1 or species 2 or species 3*. 153

Indeterminacy remains, e.g. as between fly and flying nutritious object, and as between various ways of precisifying the category *fly*. 153

Driving towards determinacy in the analogue magnitude case is the fact that the representations are used across a range of different downstream computations and behavioural outputs. 153–4

The account does not require a representation to be caused by what it represents. 154

Pietroski's case does not prove the contrary once we consider simple non-conscious systems and foreswear problematic intuitions. 154

(D) UE STRUCTURAL CORRESPONDENCE

In the rat, the relevant correspondence is with distal features, for the reason we have just seen, but there may be indeterminacy between various ways of understanding what locations are, e.g. as between absolute and relative spatial positions. 154–5

It may be indeterminate whether place cells pick out locations indexically or non-indexically. 155

(E) NATURAL PROPERTIES

The approach tends to favour natural properties. That counts against arbitrary disjunctions. 155

Such properties tend not to figure in causal explanations, which stops Peacocke-style reduced content arising in these simple systems. 155–6

More complex representations, like human conceptual representations, clearly can represent such contents, using the combinatorial power of concepts. 156

(F) DIFFERENT CONTENTS FOR DIFFERENT VEHICLES

A soft constraint deriving from the definition of UE information is that different vehicles in the same range should generally have different contents. 156

E.g. with the frog, ascribing the content *there is a fly nearby* to the firing of all retinal ganglion cells would be less explanatory of the frog's performance of its task functions. 156

My constraint: different contents for different vehicles in the same range will generally be more explanatory. 156

Soft constraint: different contents for different vehicles. 157

It follows that location is represented in the frog; numerosity (rather than *many*) in the analogue magnitude system. 157

Different components within the overall system tend to have different jobs, therefore carry different contents, although redundancy is possible. 157

(G) THE APPROPRIATE AMOUNT OF DETERMINACY

We should expect more indeterminate contents in lower level systems. 157

Components have a family of relational properties in the frog case, with not enough complexity in the system to support a way of distinguishing between them. 157

This is either indeterminacy between multiple closely related contents, or a determinate content that cannot be captured precisely in natural language. 157–8

An indeterminacy for the whole system, about a collection of UE information for a collection of internal vehicles, does not imply each content is indeterminate independently of the others, since they must fit with each other in an explanation. 158

In a system with multiple interacting components, the need for a mesh between UE information carried by different components is a significant constraint on indeterminacy. 158

(H) COMPARISON TO OTHER THEORIES

Like in Millikan, indeterminacy is reduced by appealing to causal explanation, which selects amongst coextensional properties, and by rejecting mediated explanations of stabilization. 158–9

My account adds: convergence with information; being more explicit about the constraint that different representations should have different contents; and the fact that some indeterminacy is a virtue in our subpersonal cases. 159

Papineau argues that indeterminacy is ruled out by the determinacy of desires, and by appealing to a component's specific function (following Neander). 159

My account is like Price: content is constrained by explanatory role; having an information-carrying requirement; and adopting a high church view (contra Neander and Pietroski). 159–60

Price's immediacy and abstractness conditions have roughly the same effect as my call for correlations that feature in an unmediated explanation of task function performance. With multiple components, my account also favours correlations that are specific to a component. 160

Neander is the leading proponent of low church teleosemantics, tying content to conditions that an organism can discriminate between. I don't agree that contents, or long-armed functions, are restricted to things a component is responsible for on its own. 160

Neander's second argument is based on the science of how toads manage to discriminate their prey. That practice does not imply—indeed I would argue counts against—content being tied to discriminative capacities. 160–1

My account does allow that sensorily specific properties are represented in suitably articulated systems (e.g. §4.7), but that is not the case in the (stylized) frog/toad prey-capture mechanism. 161

Ryder, Martinez, and Artiga all fix contents in terms of properties that explain statistical regularities amongst worldly conditions, in different ways. 161

Objection: the property that explains co-occurrence in the world need not be the one which explains successful behaviour. 161–2

6.3 *Compositionality and Non-Conceptual Representation*

Some features of concepts are found in the simpler systems in our case studies: semantically significant structure, unsaturated components and (limited) generality. 162

'Non-conceptual' covers all cases outside 'concepts', which are personal level, expressed in language, and constituents of beliefs and desires. 162

Concepts obey a wide-ranging generality constraint. 162

I extend 'saturated' to non-conceptual representations, without semantically significant constituent structure, that have a complete correctness and/or satisfaction condition. 162

Most of our case studies do not involve predication; although many have semantically significant structure of a simpler kind, but only limited generality. 163

The plaid motion detection system (§4.7) has different layers for colour and motion. Neither has semantically significant constituent structure. 163

The PFC colour-motion case (§4.6b) and the bee dance case do have multiple components, but the components stand alone and make claims separately. They are not unsaturated. 163

These representations do have semantically significant structure, unlike the two separate representations in the plaid motion case. 163–4

Neither the PFC case nor the bee dance case involves unsaturated elements or predication. 164

There may be unsaturated elements when place cells are used offline. 164

Our case studies do show some systematicity, and so meet a restricted, domain-specific generality constraint. 164–5

Time and place are not semantically significant features in the bee dance or PFC case (*pace* Millikan). 165

We should distinguish a different kind of 'systematicity', the kind found in a sign system exhibiting 'organization' (§5.5): there being a systematic mapping relationship from a vehicle dimension to a content dimension (e.g. more waggles = further away). 165

Concepts are reused in a wide range of contexts. Lacking that, the representations in our case studies are likely to have less determinacy. 165–6

Recap: three features of concepts are also found, to some extent, in our case studies. 166

6.4 *Objection to Relying on (Historical) Functions*

(A) SWAMPMAN

My accounts must overcome the standard challenges to relying on history in a theory of content. 166

This is made vivid by considering an intrinsic duplicate system with no history: a swampman. 166–7

All our case studies could have swamp duplicates. Why aren't the robust outcome functions of these systems enough to ground content? 167

Because, without appeal to history, there is nothing to undergird a distinction between success and failure of behaviour, in such systems. 167

E.g. consider a swamp system like the one in §4.7 that catches a ball, and another that just misses. 167

Neither would be making an error. Without stabilized functions, there is no room for a system that robustly produces an outcome in error. 168

The historical basis of stabilized functions allows us to make this distinction. Success is beneficial. 168

That intuitive argument is supported by the argument in Chapter 3 that selection, learning, and contribution to an organism's persistence are part of the cluster that gives explanatory purchase to representational content. 168–9

The stabilization process could be very recent, so a swamp system that starts interacting with its environment will rapidly acquire some task functions. 169

Other mental properties like memory would also build up rapidly in a swamp system. (Interactions with the environment would soon ground a substantial difference between Catcher and Misser.) 169

In sum, I cut down the scope of the challenge, since my view does not imply that swamp humans have no contents, and because content builds up rapidly during interactions; and then I make a positive argument that in the simple systems in our case studies content should be based on history in this way. 169

(B) COMPARISON TO MILLIKAN AND PAPINEAU

Millikan says content-based generalizations depend on the historical kind *human*, so do not carry over to swampman. 169–70

But why don't content-based generalizations run off some currently constituted category (as other generalizations about swampman do)? 170

According to varitel semantics, that is because swamp systems do not fall in the natural cluster that underpins content-based explanation. 170

Papineau says the best a posteriori reduction of content, in our world where there are no swamp systems, is historical. 170

This undershoots: the current properties are positively less explanatorily powerful because they miss the distinctive cluster, so apply more widely and support fewer inductions. 170–1

6.5 *Norms of Representation and of Function*

(A) SYSTEMATIC MISREPRESENTATION

For me, representing correctly, and well-functioning, are merely descriptive distinctions to which norms can be applied. That is all we should expect in these cases. 171

Even so I face the argument that, because fitness interests are sometimes best served by systematically misrepresenting what is the case, representing correctly cannot be equated with promoting fitness. 171–2

Peacocke's example: systematically misrepresenting the predator as closer than it is, so as to run away faster, gives a selective advantage. 172

These examples typically assume that the representation in question is involved in a second pattern of behaviour, which fixes the correct content. If so, two different contents could arise on my view. 172

Fig. 6.1 The structure of the case from Peacocke (1993). 172

Our case studies do not have that kind of structure. It has not been shown that a challenge based on systematic misrepresentation is tenable in such cases. 173

A reason to think it is not is that, without further articulation, content ends up being fixed so as to align with whatever story is told about evolutionary benefit (cp. a representation theorem in decision theory). 173

Even though I reject the possibility of radical disconnection, misrepresentation and malfunction do come apart on my view, and can do so systematically (e.g. systematic false positives). 173

(B) PSYCHOLOGICALLY PROPRIETARY REPRESENTATION

Burge (2010) makes three arguments, on normative grounds, against teleological approaches to content. Subsection (a) covered his first argument. 174

His second argument is that they are too liberal, extending to cases where content has no real explanatory value. I argue elsewhere that my contents do have explanatory value in such cases (§2.3, §8.2). 174

Burge's third argument is that contents should be distinctively psychological and normative. 174

I accept that some psychological cases are more sophisticated, but some psychological cases are covered by my approach. I argue that the requirement that representation be psychologically proprietary is not well-motivated. 174–5

On normativity, Burge has a non-reductive approach, and argues that there is no need for an account in non-semantic, non-mental, and non-normative terms. 175

I agree with Burge that such an account is not required to believe that there are representations. But such an account is more illuminating, when it is available, as I argue it is here. 175

6.6 Conclusion

Varitel semantics produces less indeterminacy than informational semantics and teleosemantics. The remaining indeterminacy is what we should expect in the systems we have been considering. 175–6

The historical component of task functions is needed just to get the explanandum—successful behaviour—into the picture. But misrepresentation is not reduced to malfunction. 176

In short, varitel semantics does a reasonable job of avoiding the standard challenges in the literature. 176

Chapter 7—Descriptive and Directive Representation

7.1 Introduction

A descriptive representation is supposed to match the world; a directive representation is supposed to make a condition obtain. 177

This chapter draws the distinction in a non-theory-neutral way, within the varitel framework. 177

'Descriptive' and 'directive' are better terms than 'indicative' and 'imperative' because the latter are used in linguistics for the grammatical mood of a sentence. 177–8

Even our simple case studies may contain other modes of representing, e.g. the suppositional. 178

With beliefs and desires, the shared condition is sometimes called the content (that p), and the mode of representing is separated out, as an attitude to that content. 178

My terminology uses 'content' more broadly, for a full specification of representational import, including mode of representing. 178

A bodily movement can also count as a 'condition' C that the organism brings about. 178

7.2 An Account of the Distinction

The accounts of content in Chapters 4 and 5 are based on exploitable relations explaining task function performance, without distinguishing between relations to inputs and relations to outputs. 179

We can supplement the accounts of content to classify the exploitable relations as playing a descriptive or directive role (or both). 179

A tempting first thought is to rely on an asymmetry in causation: caused by the world or causing an outcome. 179

But for varitel semantics the distinction should turn, not on causation, but on an asymmetry in how exploitable relations and their associated conditions figure in explaining task function performance. Descriptive representations need not be caused by their contents at all. 179

Directives are representations R where the role of the vehicle in explaining task functions depends on the fact that R produces condition C; for descriptives its explanatory role depends on C's obtaining already at the point when the behavioural outputs prompted by R occur. 179–80

The case of corollary discharge introduces a complication which means that it is easier to define directive content first. 180

Definition: directive content (based on UE information). 180

For directive content, R's role in explaining stabilization and/or robustness is to cause condition C to obtain. 180

Production of an outcome which is itself task functional can be explanatory as part of explaining how the whole system comes to produce this outcome in a way that was stabilized and robust (§4.2a). 180

Descriptive content concerns a condition C whose obtaining when R is tokened figures in explaining robustness and stabilization, but we need to exclude cases where the explanation is that R has the causal role to produce condition C. 180–1

Definition: descriptive content (based on UE information). 181

We don't want the definition to imply that all directives also have descriptive content, but we do want that to be possible (e.g. in some cases of corollary discharge). 181

Where a motor command for C has a second functional role leading to behaviour whose explanation depends on condition C obtaining independently, then it should have additional descriptive content. 181

Our definition delivers that result. 181–2

For the structural correspondence case, consider a structural representation R that has a UE structural correspondence with condition $H(x_1,x_2)$. 182

Definition: directive content (based on UE structural correspondence). 182

Definition: descriptive content (based on UE structural correspondence). 182

Applied to rat navigation, offline place cell co-activation descriptively represents that location x is near to location y. 182–3

7.3 *Application to Case Studies*

(A) UE INFORMATION

The honeybee nectar dance is a pushmi-pullyu according to my definitions of descriptive and directive content. 183

The ALCOVE model has pushmi-pullyus at output and descriptives at the input and exemplar layers. 183

Motor programs are sometimes pushmi-pullyu, with matching directive and descriptive contents. 183

The 'model state estimate' in Miall and Wolpert's (1996) predictive comparator model is in fact a pure descriptive, since it is the result of transforming the (directive) motor program into another representation. 183–4

Fig. 7.1 A predictive comparator model from Miall and Wolpert (1996). 184

The PFC colour/motion system (§4.6b) splits into some pure descriptives and some pure directives. 184–5

Pure descriptives arise in the analogue magnitude system, face recognition system and plaid motion tracking system. 185

The evidence accumulation system (§4.8) has descriptives in the cycle then directives that drive action. 185

(B) UE STRUCTURAL CORRESPONDENCE

In rat navigation, the system relies on the corresponding spatial relation obtaining. So, the correspondence has a descriptive content. 185

A directed causal graph programming a sequence of actions would have directive content. 185–6

Ryder's SINBAD model produces cells that are tuned to sources of mutual information in the inputs it has encountered. 186

Operation of the SINBAD model can be inverted so as to use descriptive representations in directive mode to drive action. 186

So far, we just have reliance on multiple exploitable correlations, both at input and at output, not reliance on structural correspondence. 186–7

However, if the network were to make use of the fact that directed connections amongst its cells correspond to causal links in the world, then we would have a case of directive UE structural correspondence. 187

It is interesting to note that this model supports something like a content–attitude distinction. Each vehicle can be deployed in either descriptive or directive mode. 187

The rat place cell system may use place cells with directive (correlational) content to drive behaviour. 187

7.4 Comparison to Existing Accounts

This section makes comparisons with three existing accounts: standard teleosemantics, decoupling, and detecting achievement of a directive. 188

Teleosemantics: R has directive content when it has the function of producing C, and descriptive content when the producer has the function of producing R when C obtains. 188

My approach falls into the same broad camp as teleosemantics. 188

Artiga argues that there will always be some, possibly very disjunctive, set of outputs that a representation R is supposed to produce; hence Millikan's view implies that all simple representation will have directive content. 188

My account does not have that consequence because disjunctive conditions do not generally qualify as content. 189

Price says a directive has to be a goal, something that is produced by a variety of means. Objection: that would exclude motor programs, which play an important role in how an organism calculates how to act. 189

Sterelny uses response breadth to pick out descriptives: they are not tightly coupled to any particular response. 189–90

Zollman distinguishes this way: descriptives are more tightly coupled to world states, directives are more tightly coupled to outputs. 190

My view tends to go along with decoupling in this way. But that is not the basis of the distinction. 190

A different view relies on deliberation: with directives, the consumer does not deliberate about how to act. 190

That is not a promising way to draw the distinction in our case studies because deliberation is not ever involved. 191

Third approach: a representation is directive when the organism can detect whether C obtains and stops pursuing C when it does. 191

This is too demanding for directive content in general, but where a system has this sophistication, my account implies that the representation is directive. 191

7.5 *Further Sophistication*

(A) MORE COMPLEX DIRECTIVE SYSTEMS

This section looks briefly at four further levels of cognitive sophistication, beyond simply separating descriptive from directive representations. 192

We saw two cases where, as with beliefs and desires, the same vehicle can be used with different directions of fit. 192

In the phenomenon of secondary conditioning, a descriptive representation of condition C has come to be a directive representation causing the organism to bring about C. 192

Detecting when you have reached your goal is a further level of sophistication. The general-purpose redeployability of every desire as a belief and every belief as a desire is a further level again. 192

Many organisms have a system for sorting amongst and prioritizing potentially conflicting directive representations, including through having directives with different (and maybe variable) strengths. 192–3

So, there are at least four levels of cognitive sophistication in which descriptive and directive representations can be embedded. 193

(B) ANOTHER MODE OF REPRESENTING

Propositional attitudes admit of other modes of representing, e.g. supposing. There may be something like that involved when place cells are active offline. 193

Offline activation of one place cell may have a content along the lines of *suppose you were at x*. Combined with the descriptive representation that *location y is near location x*, the system concludes that *y would be nearby*. 194

So, offline place cell activity is either unsaturated or suppositional. Either way, the case introduces a kind of sophistication which might have been thought to be the preserve of propositional attitudes. 194

I remain neutral as to whether the functional role described here is the same mode of representing as the propositional attitude of supposing. 194

7.6 *Conclusion*

This chapter has shown how the descriptive–directive distinction can be drawn within the varitel framework. 194–5

Chapter 8—How Content Explains

8.1 *Introduction*

Three paragraphs highlight some distinctive features of varitel semantics and introduce the chapter sections. 197–8

8.2 *How Content Explains*

(A) EXPLANATORY TRACTION IN VARITEL SEMANTICS

In order to see how content explains behaviour, the framework in Chapter 2 had contents as relational properties of real vehicles. 198–9

Now: do our accounts of content show how contents explain success and failure of behaviour? 199

For a primate in a numerosity experiment, picking the more numerous collection of objects constitutes success. 199

Contents are relational properties of components which thereby instantiate an algorithm for producing successful behaviour. 199

This explanatory practice applies to many cases—because evolution produces organisms in which robustly produced outcomes have been the target of stabilizing processes. 200

Representation arises when stabilization and robustness are achieved by internal workings (over vehicles bearing exploitable relations to distal objects and properties). 200

(B) NON-SEMANTIC CAUSAL DESCRIPTION?

Further challenge: what role is there for content, when a non-semantic causal description is always available? 200

The field of psychology looks to be full of rich content-based generalizations, but a non-semantic causal description threatens to undermine their *prima facie* explanatory force. 201

Rifle firing example: the putative semantic description marches exactly in step with the non-semantic description. 201

Varitel semantics implies that the representational explanation has vehicle–world patterns at input, and often at output, that a factorized explanation misses. 201–2

Fig. 8.1 A schematic depiction of bridging at input and output. 202

The representational explanation is based on real patterns in the way a system and its components are involved with the distal environment. 202–3

This bridging is exemplified in the analogue magnitude and rat navigation case studies. 203–4

Bodily movements considered independently of their relations to distal outcomes are often uninterpretable, e.g. the moving thumbs of someone playing a video game. 204

Bridging shows why content-based explanation breaks free from non-semantic vehicle-based explanation, allowing detailed psychological theories to have their own explanatory purchase. 204

Explanation calls for a mix of generality (breadth of application) and specificity (inductive power). Bridging delivers some generality. Specificity is offered by the rich psychological theories in which so-based contents figure. 204

(C) DOING WITHOUT TALK OF REPRESENTATION

Different challenge: why not do all the explanation in terms of correlation, correspondence, and function directly? 204–5

One version: give more fine-grained explanations, just in terms of correlation, correspondence and function. That is less explanatory. 205

Second version: accept that the clusters I point to are present and the complex properties I have constructed are important. But that is to concede everything except the term 'representation'. 205

(D) OTHER VIEWS ABOUT THE EXPLANATORY PURCHASE OF CONTENT

Ramsey argues that representational properties can earn their explanatory keep through heuristic value or causal relevance. 205–6

Egan, Shagrir, and Burge say representational contents come in to explain how an organism performs cognitive tasks. Thus, the explanandum is already given in semantic terms. 206

Dretske (1988) has contents that are causally relevant, because they are a structuring cause of behaviour. Content based on sorting behaviour (Davies) or world-involving action (Peacocke) might also be causally relevant. 206

Causal relevance might also be based on the compositionality of concepts, because that explains the systematicity of representational capacities or of behaviour. 206–7

Part of the explanatory purchase of content for varitel semantics is that content explains why the system is configured in the way it is and behaves the way it does (like Dretske 1988, but without the liberality). 207

I agree with Egan and Shagrir that contents allow us to see how a system can perform a task, although in my case the tasks are characterized non-semantically, not cognitively. 207

Vehicle properties also play a role, in showing how an algorithm works. This is a generalization of the point that compositionality of representational vehicles can explain systematicity of behaviour. 207

8.3 *Causal Efficacy of Semantic Properties*

Are semantic properties causally efficacious, or merely explanatorily relevant, on this view? 208

Varitel semantics fits Jackson and Pettit's account of why broad contents are explanatorily relevant. 208

One approach has it that both semantic properties and vehicle properties are explanatorily relevant and neither causally efficacious, with causation only in basic physics. 208

An alternative is that there can be real causal efficacy at more than one level. 208

Even if some special science properties have causal efficacy, there are further obstacles to establishing the causal efficacy of content properties. 209

My view shows why content properties are explanatorily relevant and is neutral on causal efficacy. 209

8.4 *Why Require Exploitable Relations?*

Are exploitable relations a necessary part of the content-constitution story? (Standard teleosemantics has no correlation requirement.) 209

Exploitable relations figure in my story because content is partly a matter of explaining how a system achieves its functions. 209

Output-only theories type representations by the type of behaviour produced; my account gives a reason to expect correctness conditions to obtain. 210

This means my account makes stronger predictions about the results of tokening a representation; but output-only teleosemantics can appeal to equivalent empirical generalizations, so the difference is not profound. 210

My account does a better job of showing why contents explain how a system performs its functions. 210

8.5 *Ambit of Varitel Semantics*

(A) REPRESENTATION ONLY IF CONTENT IS EXPLANATORY?

The complex properties my accounts depend on exist whether or not there is an observer able to make use of their explanatory potential. 210–11

There is no requirement for contents to be explanatorily beneficial in every case where they arise. 211

E.g. they may give little or no additional explanatory purchase in a simple thermostat. 211

Robustness comes in degrees; different stabilizing processes may be co-present and align, or not. There is less explanatory purchase in the more marginal cases. 211

Where task functions are constituted by design, the other elements may be very marginal; or these can be clear cases. 211–12

In sum: representational contents are observer-independent. The explanatory utility of the representational explanations they afford will vary. 212

Finally, is my account pragmatist? It is pragmatist in the sense that content is action-derived. But it denies that starting with action guidance implies that the role of representation should be marginalized or eliminated. 212

My emphasis on the explanatory role of content suggests pragmatism of another kind. My account certainly aims to explain the discourse of representational attribution. 212

Blackburn: the pragmatist explains by mentioning not using the referring expressions of the discourse. By those lights, varitel semantics is not pragmatist. It uses the terms 'representation', 'content', etc. (and shows what they refer to). 212–13

(B) ARE ANY CASES EXCLUDED?

Does varitel semantics imply that every natural system is a representer? 213

Simple reactions to proximal inputs, e.g. plant tropisms, do not count, because there is no robust outcome function. 213

But a plant can have representational states, e.g. if it has two ways of detecting evening, closing its flowers accordingly. 213–14

Robustness in general, e.g. in cell physiology, is not robustness in the face of different inputs, so does not in general found content. 214

Internal subpersonal cases do extend beyond the psychological, to hormonal signalling and the immune system, for example. 214

Subsystems can have robust outcome functions where the 'external' conditions are states of other parts of the organism, but these are only task functions if there is a stabilizing process (e.g. learning processes in the cell) operating below the level of the whole organism (which is where evolution by natural selection is likely to be powerful). 214–15

The functions of hormonal signalling are probably derivative from its role in serving task functions of the whole organism, not because it has task functions as a system in its own right. Similarly for the brain. 215

Although my account is more restricted than some other theories of content, it is not restricted to the psychological. 215

The kinds of content found in personal-level cases may be restricted to the psychological, but these do not cover the subpersonal-psychological. Accounts that cover the subpersonal-psychological extend more widely. 215–16

8.6 *Development and Content*

Both with concepts and with neural networks, content often concerns the circumstances in which a vehicle develops. 216

If content is just fixed by synchronic properties, it is puzzling why there should be this connection to the circumstances of development. 216

There are empirically studied cases like that. With face recognition, new vehicles arise as a result of learning/stabilizing processes and refer to their causes. 216–17

Laurence and Margolis (2002) have a theory of natural kind concepts that links the content of a new concept to the object that caused its development. 217

That connection arises, according to my accounts, because content is fixed by features of the stabilization process (task function), and because stabilization processes often give rise to representational vehicles. 217

In sum: we can see why a new representation often represents features of the things in the environment that caused it to develop. 217

8.7 *Miscellaneous Qualifications*

Can contents only be about distal objects and properties? 218

A system must have distally involved task functions, so some contents must concern distal features of the environment, but it can also have representations about proximal and internal conditions. 218

A clear example of that is meta-representation, which can arise in relatively simple systems (Shea 2014c). 218

Outputs are not restricted to bodily movements and the effects of bodily movements. Chemical and electrical outputs, for example, are included. 218–19

In the case studies, vehicles have been proper parts of a mechanism. 219

In principle, properties of the whole system could interact in content-preserving ways. 219

Functional content is a supplement to informational treatments of evolutionary signalling games. It only arises because of a stabilizing process, as do my stabilized functions. 219–20

Some game-theoretic models have meaning at attractor states that are not Nash equilibria, but as attractors they can potentially form the basis of stabilized functions, hence task functions. 220

I'm neutral about where Fregean sense or mode of presentation is needed, in addition to referential content and vehicle properties. 220

Indexicality is an important issue, but I set it aside. 220

8.8 *How to Find Out What Is Represented*

My account is a metaphysics of content, but it has implications for how we should find out what is represented. 221

Procedure: establish behaviour that has been stabilized, consider algorithms, find which one maps onto internal processes. 221

The role of task functions in specifying the explanandum is often implicit. The search for correlational information is explicit, but the restriction to information relevant to performing task functions is usually only implicit. 221

Information has to be carried in a way that is detectable by downstream processes. This is sometimes acknowledged. 221–2

There should be more emphasis on output correlations. Also on the circumstances that stabilize behaviour. 222

We can see why investigating illusions and errors is important. Ethology and comparative psychology are relevant to stabilization. 222

8.9 *Differences at the Personal Level*

How might features of the personal level make a difference to content determination? 222

Consciousness makes a difference. It could play a role in determining content; or, if it is determined by content, then various functional features of consciousness are potentially relevant to content determination (listed). 222–3

If for conscious states, or other mental states, object-level and meta-level content are fixed simultaneously, that would be relevant to content determination. 223

Relations of entailment and exclusion in a network of beliefs or concepts may play a content-constituting role. 223

For concepts, the special functional role of conscious deliberation in forming and changing our beliefs involving a concept may be relevant to content determination. 223–4

Norms applicable to belief/desire content may be interpersonal and may depend on stabilizing processes in a social group. 224

An ascriptionist theory like Dennett's intentional stance may be the right account of content for standing beliefs. 224

Will the varitel framework, at least, still be applicable? Too early to say. 224

It is wrong to think that personal-level processes could make use of subpersonal contents directly. 224–5

Pluralism amongst different kinds of personal-level representations may not be appropriate. 225

Having answered the content question for subpersonal representation should give us optimism. 225

It gives us a fixed point from which to build. Deploying our understanding of the ways in which personal-level representations are different, understanding their nature becomes a tractable research programme. 225

However, the most important achievement, if it succeeds, is that varitel semantics allows us to understand the nature of content in subpersonal representational systems. 225–6

Acknowledgements

I have been working on this topic for a long time. A lot of people have helped at various stages along the way. My most long-standing debt of gratitude is to Ruth Millikan. As philosophy started drawing me in during an MA conversion course at Birkbeck, I became convinced that intentionality was just as important an issue as consciousness or the metaphysics of mind. Ruth invited me to UConn as a visiting scholar and worked through many issues with me with great patience. I am very grateful for all the philosophical discussion, then and over the years since, and for so much warm hospitality from Ruth and her husband Don.

I moved straight on to three happy years as a PhD student at King's College London, where David Papineau was an unfailingly supportive supervisor, and subsequently a generous colleague. I'm indebted to David for helping me with early incarnations of these ideas back then, and for more recent comments on many parts of the manuscript.

The pervasive influence of Peter Godfrey-Smith's work will be apparent. Thanks to him this avid reader of his work was warmly welcomed, out of the blue, as an academic visitor to the ANU. Many of the core commitments in the book trace back to long walks around Canberra, and much of the writing has been influenced by subsequent discussions and his generous comments on talks and draft chapters.

My early years as a postdoc in Oxford were supported by a British Academy Postdoctoral Fellowship and the Mary Somerville Junior Research Fellowship at Somerville College. I am very grateful for this support, which enabled me to build up the background in psychology that underlies much of the book. Seminars and lectures I gave at Oxford and King's gave me the chance to explore this topic with students, whose questions helped to bring the issues into focus for me.

Much of the writing and some of the final research was supported by an AHRC Fellowship (AH/M005933/1), and latterly by the European Research Council (under the European Union's Horizon 2020 research and innovation programme under grant agreement No. 681422 MetCogCon); and by the fertile academic environment created by colleagues at King's and the Institute of Philosophy. Many thanks to all.

I am very grateful to the following colleagues for discussion, references, and comments on talks and draft chapters: Tim Bayne, Jonathan Birch, Colin Blakemore, Andy Clark, Martin Davies, Daniel Dennett, Nir Fresco, Ellen Fridland, Chris Frith, a reading group at Indiana University, Eva Jablonka, Kate Jeffery, John Matthewson, Gerard O'Brien, Samir Okasha, Eliot Michaelson, Matthew Parrott, Christopher Peacocke, Ron Planer, Joëlle Proust, Dan Ryder, Oron Shagrir, David Spurrett, James Stazicker, Marius Usher; and to audiences at: the Aristotelian Society; a reading group at the University of Barcelona; the Mental Representation conference at Ruhr Universität in Bochum; the British Society for the Philosophy of Science; the Centre for Philosophical Psychology University of Antwerp; the Cognitive Science seminar at CUNY Graduate Center; the European Society for Philosophy and Psychology; the Institute of Philosophy lab meetings; the International Society for the History, Philosophy, and Social Studies of Biology; the Philosophy of David Papineau conference; the Staff Seminar, Language and Cognition Seminar, and Naturalistic Approaches to the Mind Reading Group at King's College London; two LOGOS workshops in Barcelona; the Mind work-in-progress group at the University of Oxford; the Plunkett lab meeting; the Rushworth lab meeting; and the HPS seminar at Sydney University. Some of the ideas in the book received an earlier treatment in 'Naturalising Representational Content', *Philosophy Compass*, 8: 496–509 (2013); and 'Exploited Isomorphism and Structural Representation', *Proceedings of the Aristotelian Society*, 64(2): 123–44 (2014).

For detailed comments on all or most of the manuscript I'm immensely grateful to Marc Artiga, Rosa Cao, Manolo Martinez, Michael Rescorla, three readers for Oxford University Press, and, over one memorably intensive week at the IP, Patrick Butlin, Tony Cheng, Matthew Crosby, Sergio de Souza Filho, Dimitri Mollo, Kathryn Nave, and Will Sharp.

Peter Momtchiloff has been an exemplary editor at Oxford University Press. For help and advice at various stages of the publication process I'm also grateful to Molly Balikov, Hannah Chippendale, Emily Gardner, Jeremy Langworthy, Monica Matthews, Lakshmanan Sethuraman; and to Dimitri Mollo for his excellent work on the index.

Finally, it's a pleasure to be able record my gratitude to Ellie Barnes, who encouraged my swerve into the world of philosophy in the first place and has given me her unstinting support ever since.

Figure Credits

Figure 1.2 reprinted from *Current Opinion in Neurobiology*, 19 (1), Rushworth, M. S., Mars, R. B., and Summerfield, C., 'General Mechanisms for Making Decisions?' 75–83 (2009), with permission from Elsevier; figure 1.4: photograph by Warren Photographic Ltd; figure 1.5 reprinted from *Neuroscience and Biobehavioural Reviews*, 35 (2), George, O. and Koob, G. F., 'Individual Differences in Prefrontal Cortex Function and the Transition from Drug Use to Drug Dependence', 232–47 (2010), with permission from Elsevier; figure 2.1 reprinted from *Trends in Cognitive Sciences*, 2(9), Wolpert, D. M., Miall, C. R., and Kawato, M., 'Internal Models in the Cerebellum', 338–47 (1998), with permission from Elsevier; figure 2.3 and figure 1.1 used with permission from E. Barnes; figure 3.2 reprinted from *Neuropsychologia*, 36(11), Fourneret, P. and Jeannerod, M., 'Limited Conscious Monitoring of Motor Performance in Normal Subjects', 1133–40 (1998), with permission from Elsevier; figure 3.4 reprinted from *Current Opinion in Neurobiology*, 22(6), Wolpert, D. M. and Landy, M. S., 'Motor Control Is Decision Making', 996–1003 (2012), with permission from Elsevier; figure 4.2 reprinted from *Psychological Review*, 99, Kruschke, J. K., 'Alcove: An Exemplar-Based Connectionist Model of Category Learning', 22–44 (1992); figure 4.3 reprinted from *Cerebral Cortex*, 1, Felleman, D. J. and van Essen, D. C., 'Distributed Hierarchical Processing in the Primate Cerebral Cortex', 1–47 (1991), with permission from Oxford University Press and D. C. van Essen; figure 4.6 reprinted from *Nature*, 503 (7474), Mante, V., Sussillo, D., Shenoy, K. D., and Newsome, W. T., 'Context-Dependent Computation by Recurrent Dynamics in Prefrontal Cortex', 78–84 (2013), with permission from Springer Nature; figure 4.9 reprinted from *Neuron*, 13 (1), van Essen, D. C. and Gallant, J. L., 'Neural Mechanisms of Form and Motion Processing in the Primate Visual System', 1–10 (1994), with permission from Elsevier; figure 4.10 reprinted/adapted from 'Optimal Decision Making in the Cortico-Basal-Ganglia Circuit', Bogacz, R., in: Forstmann, B. and Wagenmakers, E. J. (eds.), *An Introduction to Model-Based Cognitive Neuroscience*, 291–302 (2015), with permission from Springer Nature; figure 5.2 reprinted by permission from Springer Nature: *Nature*, 416 (6876), Lever, C., Wills, T., Cacucci, F., Burgess, N., and O'Keefe, J., 'Long-Term Plasticity in Hippocampal Place-Cell Representation of Environmental Geometry', 90–4 (2002), as reproduced in *Hippocampus*, 15(7), O'Keefe, J. and Burgess, N., 'Dual Phase and Rate Coding in Hippocampal Place Cells: Theoretical Significance and Relationship to Entorhinal Grid Cells', 853–66 (2005); figures 5.4 and 5.5 used with permission from Robert Cummins; figure 5.7 reprinted from *Proceedings of the National Academy of Sciences*, 112, Huys, Q. J. M., Lally, N., Faulkner, P., Eshel, N., Seifritz, E., Gershman, S. J., Dayan, P., and Roiser, J. P., 'Interplay of Approximate Planning Strategies', 3098–103 (2015); figure 7.1 reprinted from *Neural Networks*, 9(8), Miall, R. C. and Wolpert, D. M., 'Forward Models for Physiological Motor Control', 1265–79 (1996), with permission from Elsevier.

References

Each reference is followed by the page numbers of the pages where that work is cited in the book

Abell, F., F. Happe, and U. Frith. 2000. 'Do Triangles Play Tricks? Attribution of Mental States to Animated Shapes in Normal and Abnormal Development', *Cognitive Development*, 15: 1–15. 52

Adelson, Edward H., and J. Anthony Movshon. 1982. 'Phenomenal Coherence of Moving Visual Patterns', *Nature*, 300: 523–5. 104

Alexander, W. H., and J. W. Brown. 2011. 'Medial Prefrontal Cortex as an Action-Outcome Predictor', *Nature Neuroscience*, 14: 1338–44. 87

Alon, Uri, Michael G. Surette, Naama Barkai, and Stanislas Leibler. 1999. 'Robustness in Bacterial Chemotaxis', *Nature*, 397: 168–71. 59

Anderson, Michael, and Anthony Chemero. 2016. 'The Brain Evolved to Guide Action'. In Shepherd, ed., *The Wiley Handbook of Evolutionary Neuroscience*. Chichester: John Wiley & Sons, 1–20. 212

Andrade, Maydianne. 1996. 'Sexual Selection for Male Sacrifice in Redback Spiders', *Science*, 271: 70–2. 72

Apperly, I. A., and S. A. Butterfill. 2009. 'Do Humans Have Two Systems to Track Beliefs and Belief-Like States?', *Psychological Review*, 116: 953. 8

Artiga, Marc. 2014a. 'Teleosemantics and Pushmi-Pullyu Representations', *Erkenntnis*, 79: 545–66. 188–9, 191

Artiga, Marc. 2014b. 'The Modal Theory of Function Is Not About Functions', *Philosophy of Science*, 81: 580–91. 73

Artiga, Marc. 2016. 'Teleosemantic Modeling of Cognitive Representations', *Biology & Philosophy*, 31: 483–505. 95

Artiga, Marc. In submission. 'Beyond Black Dots and Nutritious Things: A Solution to the Indeterminacy Problem'. 161

Artiga, Marc, and Manolo Martinez. 2016. 'The Organizational Account of Function Is an Etiological Account of Function', *Acta Biotheoretica*, 64: 1–13. 72

Aschersleben, Gisa, Tanja Hofer, and Bianca Jovanovic. 2008. 'The Link between Infant Attention to Goal-Directed Action and Later Theory of Mind Abilities', *Developmental Science*, 11: 862–8. 52

Balaguer, Jan, Hugo Spiers, Demis Hassabis, and Christopher Summerfield. 2016. 'Neural Mechanisms of Hierarchical Planning in a Virtual Subway Network', *Neuron*, 90: 893–903. 135

Barrett, Louise. 2011. *Beyond the Brain: How Body and Environment Shape Animal and Human Minds*. Princeton University Press. 212

Barth, H., N. Kanwisher, and E. Spelke. 2003. 'The Construction of Large Number Representations in Adults', *Cognition*, 86: 201–21. 98

Bastian, Amy J. 2006. 'Learning to Predict the Future: The Cerebellum Adapts Feedforward Movement Control', *Current Opinion in Neurobiology*, 16: 645–9. 53, 67

Battaglia-Mayer, Alexandra, Tania Buiatti, Roberto Caminiti, Stefano Ferraina, Francesco Lacquaniti, and Tim Shallice. 2014. 'Correction and Suppression of Reaching Movements in the Cerebral Cortex: Physiological and Neuropsychological Aspects', *Neuroscience & Biobehavioral Reviews*, 42: 232–51. 67

Bedau, Mark. 1992. 'Goal-Directed Systems and the Good', *The Monist*, 75: 34–51. 54

Bell, Andrew H., Tatiana Pasternak, and Leslie G. Ungerleider. 2014. 'Ventral and Dorsal Cortical Processing Streams'. In Werner and Chalupa, eds., *The New Visual Neurosciences*. Cambridge, MA: MIT Press, 226–41. 104

Bellmund, Jacob L. S., Lorena Deuker, Tobias Navarro Schröder, and Christian F. Doeller. 2016. 'Grid-Cell Representations in Mental Simulation', *Elife*, 5: e17089. 115

Bennett, Karen. 2003. 'Why the Exclusion Problem Seems Intractable, and How, Just Maybe, to Tract It', *Nous*, 37: 471–97. 208

Berg, Howard C., and Douglas A. Brown. 1972. 'Chemotaxis in Escherichia Coli Analysed by Three-Dimensional Tracking', *Nature*, 239: 500–4. 58

Bigelow, John, and Robert Pargetter. 1987. 'Functions', *Journal of Philosophy*, 84: 181–96. 72–3

Biro, Szilvia, and Alan M. Leslie. 2007. 'Infants' Perception of Goal-Directed Actions: Development through Cue-Based Bootstrapping', *Developmental Science*, 10: 379–98. 52

Blackburn, Simon. 2010. 'The Steps from Doing to Saying', *Proceedings of the Aristotelian Society*, 110: 1–13. 211

Block, Ned. 1986. 'Advertisement for a Semantics for Psychology'. In French, Uehling and Wettstein, eds., *Midwest Studies in Philosophy*, X: *Studies in the Philosophy of Mind*. Minneapolis: University of Minnesota Press, 615–78. 13

Blumson, Ben. 2012. 'Mental Maps', *Philosophy and Phenomenological Research*, 85: 413–34. 125

Bogacz, Rafal. 2015. 'Optimal Decision Making in the Cortico-Basal-Ganglia Circuit'. In Forstmann and Wagenmakers, eds., *An Introduction to Model-Based Cognitive Neuroscience*. New York: Springer, 291–302. 106, 107, 109

Bontley, Tom. 1998. 'Individualism and the Nature of Syntactic States', *British Journal for the Philosophy of Science*, 49: 557–74. 40

Boorse, Christopher. 1976. 'Wright on Functions', *Philosophical Review*, 85: 70–86. 57

Bouisset, S., and M. Zattara. 1981. 'A Sequence of Postural Movements Precedes Voluntary Movement', *Neuroscience letters*, 22: 263–70. 181

Boyd, R. 1991. 'Realism, Anti-Foundationalism and the Enthusiasm for Natural Kinds', *Philosophical Studies*, 61: 127–48. 51

Bradley, A. J., I. R. McDonald, and A. K. Lee. 1980. 'Stress and Mortality in a Small Marsupial (Antechinus Stuartii, Macleay)', *General and Comparative Endocrinology*, 40: 188–200. 72

Braithwaite, R. B. 1933. 'The Nature of Believing', *Proceedings of the Aristotelian Society*, 33: 129–46. 16

Brannon, Elizabeth M., and Herbert S. Terrace. 1998. 'Ordering of the Numerosities 1 to 9 by Monkeys', *Science*, 282: 746–9. 98

Brentano, F. C. 1874/1995. *Psychology from an Empirical Standpoint*. London: Routledge. 8

Burge, Tyler. 2010. *Origins of Objectivity*. Oxford University Press. 174–5, 202, 206, 218

Burr, David. 2014. 'Motion Perception: Human Psychophysics'. In Werner and Chalupa, eds., *The New Visual Neurosciences*. Cambridge, MA: MIT Press, 763–75. 104

Byrne, Alex. 2005. 'Perception and Conceptual Content'. In Sosa and Steup, eds., *Contemporary Debates in Epistemology*. Oxford: Blackwell, 231–50. 162

Camp, Elisabeth. 2007. 'Thinking with Maps', *Philosophical Perspectives*, 21: 145–82. 125

Camp, Elisabeth. 2009. 'Putting Thoughts to Work: Concepts, Systematicity, and Stimulus-Independence', *Philosophy and Phenomenological Research*, 78: 275–311. 115, 206

Cao, Rosa. 2012. 'Teleosemantic Approaches to Information in the Brain', *Biology & Philosophy*, 27: 49–71. 19, 95

Cao, Rosa. 2014. 'Signaling in the Brain', *Philosophy of Science*, 81: 891–901. 95

Carey, Susan. 2009. *The Origin of Concepts*. Oxford University Press. 13, 98, 126

Carruthers, Peter. 2011. *The Opacity of Mind: An Integrative Theory of Self-Knowledge*. Oxford University Press. 28

Charest, Ian, Rogier A. Kievit, Taylor W. Schmitz, Diana Deca, and Nikolaus Kriegeskorte. 2014. 'Unique Semantic Space in the Brain of Each Beholder Predicts Perceived Similarity', *Proceedings of the National Academy of Sciences*, 111: 14565–70. 133

Chestek, Cynthia A., Aaron P. Batista, Gopal Santhanam, M. Yu Byron, Afsheen Afshar, John P. Cunningham, Vikash Gilja, Stephen I. Ryu, Mark M. Churchland, and Krishna V. Shenoy. 2007. 'Single-Neuron Stability During Repeated Reaching in Macaque Premotor Cortex', *Journal of Neuroscience*, 27: 10742–50. 59

Chklovskii, Dmitri B., and Alexei A. Koulakov. 2004. 'Maps in the Brain: What Can We Learn from Them?', *Annual Review of Neuroscience*, 27: 369–92. 120

Christensen, Wayne D., and Mark H. Bickhard. 2002. 'The Process Dynamics of Normative Function', *The Monist*, 85: 3–28. 58, 63

Churchland, Paul M. 1998. 'Conceptual Similarity across Sensory and Neural Diversity: The Fodor/Lepore Challenge Answered', *Journal of Philosophy*, 95: 5–32. 13–14, 132

Churchland, Paul M. 2012. *Plato's Camera: How the Physical Brain Captures a Landscape of Abstract Universals*. London/Cambridge, MA: MIT Press. 13–14, 132

Clower, Dottie M., John M. Hoffman, John R. Votaw, Tracy L. Faber, Roger P. Woods, and Garrett E. Alexander. 1996. 'Role of Posterior Parietal Cortex in the Recalibration of Visually Guided Reaching', *Nature*, 383: 618–21. 53

Cohen, Jonathan D., and Frank Tong. 2001. 'The Face of Controversy', *Science*, 293: 2405–7. 216

Colwill, Ruth M., and Robert A. Rescorla. 1988. 'Associations between the Discriminative Stimulus and the Reinforcer in Instrumental Learning', *Journal of Experimental Psychology: Animal Behavior Processes*, 14: 155. 192

Constantinescu, Alexandra O., Jill X. O'Reilly, and Timothy E. J. Behrens. 2016. 'Organizing Conceptual Knowledge in Humans with a Gridlike Code', *Science*, 352: 1464–8. 133, 139, 143

Corneil, Dane S., and Wulfram Gerstner. 2015. 'Attractor Network Dynamics Enable Preplay and Rapid Path Planning in Maze-Like Environments'. In C. Cortes, N. D. Lawrence, D. D. Lee, M. Sugiyama and R. Garnett, eds., *Advances in Neural Information Processing Systems* 28. New York: Curran Associates, Inc., 1684–92. 115

Corrado, G. S., L. P. Sugrue, J. R. Brown, and W. T. Newsome. 2009. 'The Trouble with Choice: Studying Decision Variables in the Brain'. In Glimcher, Camerer, Fehr, and Poldrack, eds., *Neuroeconomics: Decision Making and the Brain*. Amsterdam: Elsevier, 463–80. 85

Cover, Thomas M., and Joy A. Thomas. 2006. *Elements of Information Theory*. 2nd edn. Hoboken, NJ: John Wiley & Sons. 12

Crane, Tim. 1990. 'The Language of Thought: No Syntax without Semantics', *Mind & Language*, 5: 187–212. 40

Crapse, Trinity B., and Marc A. Sommer. 2008. 'Corollary Discharge across the Animal Kingdom', *Nature Reviews Neuroscience*, 9: 587–600. 97

Craver, Carl F. 2013. 'Functions and Mechanisms: A Perspectivalist View'. In Huneman, ed., *Functions: Selection and Mechanisms*. London/New York: Springer, 133–58. 203

Craver, Carl F. 2014. 'The Ontic Account of Scientific Explanation'. In Kaiser, Scholz, Plenge and Hüttemann, eds., *Explanation in the Special Sciences: The Case of Biology and History*. Dordrecht: Springer, 27–52. 88

Croner, Lisa J., and Thomas D. Albright. 1999. 'Segmentation by Color Influences Responses of Motion-Sensitive Neurons in the Cortical Middle Temporal Visual Area', *Journal of Neuroscience*, 19: 3935–51. 104

Cummins, Robert. 1984. 'Functional Analysis'. In Sober, ed., *Conceptual Issues in Evolutionary Biology: An Anthology*. Cambridge, MA: Bradford, MIT Press. 51

Cummins, Robert. 1989. *Meaning and Mental Representation*. Cambridge, MA: MIT Press. 14, 51, 112

Cummins, Robert. 1996. *Representations, Targets, and Attitudes*. Cambridge, MA: Bradford, MIT Press. 51, 128, 129

Cummins, Robert, Jim Blackmon, David Byrd, Alexa Lee, and Martin Roth. 2006. 'Representation and Unexploited Content'. In MacDonald and Papineau, eds., *Teleosemantics*. Oxford University Press. 203

Danks, David. 2014. *Unifying the Mind: Cognitive Representations as Graphical Models*. London/Cambridge MA: MIT Press. 13

Davidson, Donald. 1974a. 'Psychology as Philosophy'. In Brown, ed., *Philosophy of Psychology*. London: Macmillan, 41–52. 14

Davidson, Donald. 1974b. 'Belief and the Basis of Meaning', *Synthese*, 27: 309–23. 14

Davies, Martin. 1991. 'Individualism and Perceptual Content', *Mind*, 100: 461–84. 205

Davies, Martin. 2005. 'An Approach to Philosophy of Cognitive Science'. In Jackson and Smith, eds., *The Oxford Handbook of Contemporary Philosophy*. Oxford University Press. vi

Daw, Nathaniel D., Samuel J. Gershman, Ben Seymour, Peter Dayan, and Raymond J. Dolan. 2011. 'Model-Based Influences on Humans' Choices and Striatal Prediction Errors', *Neuron*, 69: 1204–15. 135

Daw, Nathaniel D., and Peter Dayan. 2014. 'The Algorithmic Anatomy of Model-Based Evaluation', *Philosophical Transactions of the Royal Society B*, 369: 20130478. 135

Dayan, Peter. 2014. 'Rationalizable Irrationalities of Choice', *Topics in Cognitive Science*, 6: 204–28. 134

De Almeida, Licurgo, Marco Idiart, Aline Villavicencio, and John Lisman. 2012. 'Alternating Predictive and Short-Term Memory Modes of Entorhinal Grid Cells', *Hippocampus*, 22: 1647–51. 115

Deadwyler, Sam A., Terence Bunn, and Robert E. Hampson. 1996. 'Hippocampal Ensemble Activity During Spatial Delayed-Nonmatch-to-Sample Performance in Rats', *Journal of Neuroscience*, 16: 354–72. 114

deCharms, R. C., and A. Zador. 2000. 'Neural Representation and the Cortical Code', *Annual Review of Neuroscience*, 23: 613–47. 80

Dehaene, S. 1997. *The Number Sense*. Oxford University Press. 98

Dennett, Daniel C. 1971. 'Intentional Systems', *Journal of Philosophy*, 68: 87–106. 31

Dennett, Daniel C. 1978. 'Artificial Intelligence as Philosophy and as Psychology'. In *Brainstorms: Philosophical Essays on Mind and Psychology*. Cambridge, MA: MIT Press, 109–26. 36

Dennett, Daniel C. 1981. 'True Believers: The Intentional Strategy and Why It Works'. In Heath, ed., *Scientific Explanation*. Oxford University Press, 53–76. 14, 224

Dennett, Daniel C. 1991. 'Real Patterns', *Journal of Philosophy*, 88: 27–51. 14, 32, 203

Descartes, R. 1637/1988. *Selected Philosophical Writings*. Ed. J. Cottingham. Cambridge University Press. 3

Desmurget, Michel, and Scott Grafton. 2000. 'Forward Modeling Allows Feedback Control for Fast Reaching Movements', *Trends in Cognitive Sciences*, 4: 423–31. 67, 183

Diamond, Jared M. 1982. 'Big-Bang Reproduction and Ageing in Male Marsupial Mice', *Nature*, 298: 115–16. 72

Diba, Kamran, and György Buzsáki. 2007. 'Forward and Reverse Hippocampal Place-Cell Sequences During Ripples', *Nature Neuroscience*, 10: 1241–2. 115

Dickie, Imogen. 2015. *Fixing Reference*. Oxford University Press. 191, 192

Dragoi, George, and Susumu Tonegawa. 2011. 'Preplay of Future Place Cell Sequences by Hippocampal Cellular Assemblies', *Nature*, 469: 397–401. 115

Dragoi, George, and Susumu Tonegawa. 2013. 'Distinct Preplay of Multiple Novel Spatial Experiences in the Rat', *Proceedings of the National Academy of Sciences*, 110: 9100–5. 115

Dretske, Fred. 1981. *Knowledge and the Flow of Information*. Cambridge, MA: MIT Press. 12

Dretske, Fred. 1986. 'Misrepresentation'. In Bogdan, ed., *Belief: Form, Content and Function*. Oxford University Press. 30, 87, 203

Dretske, Fred. 1988. *Explaining Behaviour: Reasons in a World of Causes*. Cambridge, MA: MIT Press. 21, 22, 23, 42, 87, 160, 192, 206, 207

Dretske, Fred. 1991. 'Dretske's Replies'. In McLaughlin, ed., *Dretske and His Critics*. Oxford: Blackwell, 180–221. 21, 87, 207

Edin, Benoni B. 2008. 'Assigning Biological Functions: Making Sense of Causal Chains', *Synthese*, 161: 203–18. 58

Egan, Frances. 1991. 'Must Psychology Be Individualistic', *The Philosophical Review*, 100: 179–203. 35

Egan, Frances. 2014. 'How to Think About Mental Content', *Philosophical Studies*, 170: 115–35. 205, 206

Eliasmith, Chris. 2010. 'How We Ought to Describe Computation in the Brain', *Studies in History and Philosophy of Science Part A*, 41: 313–20. 34

Eliasmith, Chris. 2013. *How to Build a Brain: A Neural Architecture for Biological Cognition*. Oxford University Press. 12, 185–6

Essen, D. C. van, and J. L. Gallant. 1994. 'Neural Mechanisms of Form and Motion Processing in the Primate Visual System', *Neuron*, 13: 1–10. 103–5

Evans, Gareth. 1982. *The Varieties of Reference*. Oxford University Press.

Felleman, Daniel J., and David C. van Essen. 1991. 'Distributed Hierarchical Processing in the Primate Cerebral Cortex', *Cerebral Cortex*, 1: 1–47. 94

Fodor, Jerry A. 1974. 'Special Sciences, or the Disunity of Science as a Working Hypothesis', *Synthese*, 28: 97–115. 26

Fodor, Jerry A. 1975. *The Language of Thought*. Cambridge, MA: Harvard University Press. 207

Fodor, Jerry A. 1987a. *Psychosemantics*. Cambridge, MA: MIT Press. 26, 178

Fodor, Jerry A. 1987b. 'Why There Still Has to Be a Language of Thought'. In *Psychosemantics*. Cambridge, MA: MIT Press. 164, 207

Fodor, Jerry A. 1990. *A Theory of Content and Other Essays*. Cambridge, MA: MIT Press. 150

Fodor, Jerry A. 1991. 'Hedged Laws and Psychological Explanations', *Mind*, 100: 19–33. 23

Fodor, Jerry A. 2008. *LOT* 2. Oxford University Press. 216

Fodor, Jerry A., and E. Lepore. 1992. *Holism: A Shopper's Guide*. Oxford: Wiley-Blackwell. 13

Fodor, Jerry A., and Zenon W. Pylyshyn. 1988. 'Connectionism and Cognitive Architecture: A Critical Analysis', *Cognition*, 28: 3–71. 206

Forster, L. M. 1992. 'The Stereotyped Behavior of Sexual Cannibalism in Latrodectus-Hasselti Thorell (Araneae, Theridiidae), the Australian Redback Spider', *Australian Journal of Zoology*, 40: 1–11. 72

Foster, David J., and Matthew A. Wilson. 2006. 'Reverse Replay of Behavioural Sequences in Hippocampal Place Cells During the Awake State', *Nature*, 440: 680–3. 115

Fourneret, Pierre, and Marc Jeannerod. 1998. 'Limited Conscious Monitoring of Motor Performance in Normal Subjects', *Neuropsychologia*, 36: 1133–40. 53

Franklin, David W., and Daniel M. Wolpert. 2011. 'Computational Mechanisms of Sensorimotor Control', *Neuron*, 72: 425–42. 25, 26

Frith, Chris D., and Uta Frith. 1999. 'Interacting Minds: A Biological Basis', *Science*, 286: 1692–5. 52

Gallese, Vittorio, Luciano Fadiga, Leonardo Fogassi, and Giacomo Rizzolatti. 1996. 'Action Recognition in the Premotor Cortex', *Brain*, 119: 593–609. 8

Gallistel, C. R. 1990. *The Organization of Learning*. London/Cambridge MA: MIT Press. 130–1

George, Olivier, and George F. Koob. 2010. 'Individual Differences in Prefrontal Cortex Function and the Transition from Drug Use to Drug Dependence', *Neuroscience and Biobehavioral Reviews*, 35: 232–47. 20

Gergely, Gyorgy, and Gergely Csibra. 2003. 'Teleological Reasoning in Infancy: The Naive Theory of Rational Action', *Trends in Cognitive Sciences*, 7: 287–92. 52

Gläscher, Jan, Nathaniel Daw, Peter Dayan, and John P. O'Doherty. 2010. 'States Versus Rewards: Dissociable Neural Prediction Error Signals Underlying Model-Based and Model-Free Reinforcement Learning', *Neuron*, 66: 585–95. 135

Godfrey-Smith, Peter. 1989. 'Misinformation', *Canadian Journal of Philosophy*, 19: 533–50. 141, 173

Godfrey-Smith, Peter. 1991. 'Signal, Decision, Action', *Journal of Philosophy*, 88: 709–22. 143, 173

Godfrey-Smith, Peter. 1992. 'Indication and Adaptation', *Synthese*, 92: 283–312. 206, 207

Godfrey-Smith, Peter. 1994a. 'A Continuum of Semantic Optimism'. In Stich and Warfield, eds., *Mental Representation: A Reader*. Oxford: Blackwell, 259–77. 14, 150, 159

Godfrey-Smith, Peter. 1994b. 'A Modern History Theory of Functions', *Nous*, 28: 344–62. 63, 73

Godfrey-Smith, Peter. 1996. *Complexity and the Function of Mind in Nature*. Cambridge University Press. 112, 210

Godfrey-Smith, Peter. 2004. 'On Folk Psychology and Mental Representation'. In Clapin, Staines and Slezak, eds., *Representation in Mind: New Approaches to Mental Representation*. Amsterdam: Elsevier, 147–62. 42

Godfrey-Smith, Peter. 2006. 'Mental Representation, Naturalism and Teleosemantics'. In Papineau and Macdonald, eds., *New Essays on Teleosemantics*. Oxford University Press, 42–68. 15, 35, 197

Godfrey-Smith, Peter. 2008. 'Explanation in Evolutionary Biology: Comments on Fodor', *Mind & Language*, 23: 32–41. 150

Godfrey-Smith, Peter. 2013. 'Signals, Icons, and Beliefs'. In Ryder, Kingsbury, and Williford, eds., *Millikan and Her Critics*. Oxford/Malden MA: Wiley-Blackwell, 41–58. 95, 115

Godfrey-Smith, Peter. 2016. 'Individuality, Subjectivity, and Minimal Cognition', *Biology & Philosophy*, 31: 775–96. 58

Godfrey-Smith, Peter. 2017. 'Senders, Receivers, and Symbolic Artifacts', *Biological Theory*, 12: 275–86. 127, 165

Goodale, Melvyn A., Denis Pelisson, and Claude Prablanc. 1986. 'Large Adjustments in Visually Guided Reaching Do Not Depend on Vision of the Hand or Perception of Target Displacement', *Nature*, 320: 748. 53

Goodman, Nelson. 1972. 'Seven Strictures on Similarity'. *Problems and Projects*. New York: Bobbs-Merrill, 437–46. 112

Goodman, Noah D., Vikash K. Mansinghka, and Joshua B. Tenenbaum. 2007. 'Learning Grounded Causal Models'. In D. S. McNamara, and J. G. Trafton, eds., *Proceedings of the 29th Annual Conference of the Cognitive Science Society*. Austin, TX: Cognitive Science Society, 305–10. 135

Gopnik, Alison, and Henry M. Wellman. 2012. 'Reconstructing Constructivism: Causal Models, Bayesian Learning Mechanisms, and the Theory Theory', *Psychological Bulletin*, 138: 1085. 13

Griffiths, Paul Edmund. 2009. 'In What Sense Does "Nothing Make Sense Except in the Light of Evolution"?', *Acta Biotheoretica*, 57: 11–32. 49, 72

Horner, Aidan J., James A. Bisby, Ewa Zotow, Daniel Bush, and Neil Burgess. 2016. 'Grid-Like Processing of Imagined Navigation', *Current Biology*, 26: 842–7. 115

Hornsby, Jennifer. 1997. *Simple Mindedness: A Defence of Naïve Naturalism in the Philosophy of Mind*. Cambridge, MA: Harvard University Press. 26

Hornsby, Jennifer. 2000. 'Personal and Sub-Personal: A Defence of Dennett's Early Distinction', *Philosophical Explorations*, 3: 6–24. 26

Horowitz, A. 2007. 'Computation, External Factors, and Cognitive Explanations', *Philosophical Psychology*, 20: 65–80. 40

Hsieh, Liang-Tien, Matthias J. Gruber, Lucas J. Jenkins, and Charan Ranganath. 2014. 'Hippocampal Activity Patterns Carry Information about Objects in Temporal Context', *Neuron*, 81: 1165–78. 136

Hsieh, Yi-Ju, and Barry L. Wanner. 2010. 'Global Regulation by the Seven-Component P I Signaling System', *Current Opinion in Microbiology*, 13: 198–203. 213

Hubel, David H., and Torsten N. Wiesel. 1962. 'Receptive Fields, Binocular Interaction and Functional Architecture in the Cat's Visual Cortex', *Journal of Physiology*, 160: 106–54. 80

Humberstone, I. Lloyd. 1992. 'Direction of Fit', *Mind*, 101: 59–83. 177

Hummel, J. E., and I. Biederman. 1992. 'Dynamic Binding in a Neural Network for Shape Recognition', *Psychological Review*, 99: 480–517. 93

Hunt, L. T., N. Kolling, A. Soltani, M. W. Woolrich, M. F. Rushworth, and T. E. Behrens. 2012. 'Mechanisms Underlying Cortical Activity During Value-Guided Choice', *Nature Neuroscience*, 15: 470–6, S1-3. 80, 221

Huth, Alexander G., Shinji Nishimoto, An T. Vu, and Jack L. Gallant. 2012. 'A Continuous Semantic Space Describes the Representation of Thousands of Object and Action Categories across the Human Brain', *Neuron*, 76: 1210–24. 133

Huttegger, Simon M. 2007. 'Evolutionary Explanations of Indicatives and Imperatives', *Erkenntnis*, 66: 409–36. 190

Hutto, Daniel D. and Glenda Satne. 2015. 'The Natural Origins of Content', *Philosophia*, 43: 521–36. 205, 212

Huys, Quentin J. M., Neir Eshel, Elizabeth O'Nions, Luke Sheridan, Peter Dayan, and Jonathan P. Roiser. 2012. 'Bonsai Trees in Your Head: How the Pavlovian System Sculpts Goal-Directed Choices by Pruning Decision Trees', *PLoS Computational Biology*, 8: e1002410. 135, 136

Huys, Quentin J. M., Níall Lally, Paul Faulkner, Neir Eshel, Erich Seifritz, Samuel J. Gershman, Peter Dayan, and Jonathan P. Roiser. 2015. 'Interplay of Approximate Planning Strategies', *Proceedings of the National Academy of Sciences*, 112: 3098–103. 135, 136

Jackson, Frank, and Philip Pettit. 1988. 'Functionalism and Broad Content', *Mind*, 97: 381–400. 208

Jackson, Frank, and Philip Pettit. 1990. 'Program Explanation: A General Perspective', *Analysis*, 50: 107–17. 208

Johansson, Petter, Lars Hall, Sverker Sikström, and Andreas Olsson. 2005. 'Failure to Detect Mismatches between Intention and Outcome in a Simple Decision Task', *Science*, 310: 116–19. 28

Johnson, Mark H., Suzanne Dziurawiec, Hadyn Ellis, and John Morton. 1991. 'Newborns' Preferential Tracking of Face-Like Stimuli and Its Subsequent Decline', *Cognition*, 40: 1–19. 60

Kanwisher, Nancy. 2000. 'Domain Specificity in Face Perception', *Nature Neuroscience*, 3: 759. 216

Katz, L. N., J. L. Yates, J. W. Pillow, and A. C. Huk. 2016. 'Dissociated Functional Significance of Decision-Related Activity in the Primate Dorsal Stream', *Nature*, 535: 285–8. 221

Khajeh-Alijani, Azadeh, Robert Urbanczik, and Walter Senn. 2015. 'Scale-Free Navigational Planning by Neuronal Traveling Waves', *PLOS One*, 10: e0127269. 115

Kiani, Roozbeh, Hossein Esteky, Koorosh Mirpour, and Keiji Tanaka. 2007. 'Object Category Structure in Response Patterns of Neuronal Population in Monkey Inferior Temporal Cortex', *Journal of Neurophysiology*, 97: 4296–309. 133

Kiani, Roozbeh, and Michael N. Shadlen. 2009. 'Representation of Confidence Associated with a Decision by Neurons in the Parietal Cortex', *Science*, 324: 759–64. 59

Kingsbury, J. 2008. 'Learning and Selection', *Biology & Philosophy*, 23: 493–507. 21

Kirschner, Marc, and John Gerhart. 1998. 'Evolvability', *Proceedings of the National Academy of Sciences*, 95: 8420–7. 214

Knudsen, Eric I., S. du Lac, and Steven D. Esterly. 1987. 'Computational Maps in the Brain', *Annual Review of Neuroscience*, 10: 41–65. 120

Koechlin, Etienne, and Alexandre Hyafil. 2007. 'Anterior Prefrontal Function and the Limits of Human Decision-Making', *Science*, 318: 594–8. 135

Koechlin, Etienne, C. Ody, and F. Kouneiher. 2003. 'The Architecture of Cognitive Control in the Human Prefrontal Cortex', *Science*, 302: 1181–5. 135

Krasensky, Julia, and Claudia Jonak. 2012. 'Drought, Salt, and Temperature Stress-Induced Metabolic Rearrangements and Regulatory Networks', *Journal of Experimental Botany*, 63: 1593–608. 214

Kriegeskorte, Nikolaus. 2015. 'Deep Neural Networks: A New Framework for Modeling Biological Vision and Brain Information Processing', *Annual Review of Vision Science*, 1: 417–46. 91

Krizhevsky, Alex, Ilya Sutskever, and Geoffrey E. Hinton. 2012. 'Imagenet Classification with Deep Convolutional Neural Networks'. In F. Pereira, C. J. C. Burges, L. Bottou and K. Q. Weinberger, eds., *Advances in Neural Information Processing Systems* 25. New York: Curran Associates, Inc., 1097–105. 91

Kropff, Emilio, James E. Carmichael, May-Britt Moser, and Edvard I. Moser. 2015. 'Speed Cells in the Medial Entorhinal Cortex', *Nature*, 523: 419–24. 115

Kruschke, J. K. 1992. 'Alcove: An Exemplar-Based Connectionist Model of Category Learning', *Psychological Review*, 99: 22–44. 91

Kurth-Nelson, Zeb, Marcos Economides, Raymond J. Dolan, and Peter Dayan. 2016. 'Fast Sequences of Non-Spatial State Representations in Humans', *Neuron*, 91: 194–204. 136

Ladyman, James. 2017. 'An Apology for Naturalized Metaphysics'. In Slater and Yudell, eds., *Metaphysics and the Philosophy of Science: New Essays*. Oxford University Press, 141. 32

Ladyman, James, and Don Ross. 2007. *Every Thing Must Go*. Oxford University Press. 32, 203

Laurence, Stephen, and Eric Margolis. 2002. 'Radical Concept Nativism', *Cognition*, 86: 25–55. 217

Lewis, David. 1969. *Convention*. Cambridge, MA: Harvard University Press. 190, 219

Love, B. C., D. L. Medin, and T. M. Gureckis. 2004. 'Sustain: A Network Model of Category Learning', *Psychological Review*, 111: 309–32. 93

Lyon, Pamela. 2017. 'Environmental Complexity, Adaptability and Bacterial Cognition: Godfrey-Smith's Hypothesis under the Microscope', *Biology & Philosophy*, 32: 443–65. 213

Mante, Valerio, David Sussillo, Krishna V. Shenoy, and William T. Newsome. 2013. 'Context-Dependent Computation by Recurrent Dynamics in Prefrontal Cortex', *Nature*, 503: 78–84. 100, 101, 102, 184

Marr, D. 1982. *Vision*. New York: W. H. Freeman & Co. 34, 137

Mars, R. B., Nicholas Shea, N. Kolling, and M. F. S. Rushworth. 2012. 'Model-Based Analyses: Promises, Pitfalls, and Example Applications to the Study of Cognitive Control', *Quarterly Journal of Experimental Psychology*, 65: 252–67. 85

Martin, Eugene V. Koonin William. 2005. 'On the Origin of Genomes and Cells within Inorganic Compartments', *Trends in Genetics*, 21: 647–53. 58

Martínez, Manolo. 2013. 'Teleosemantics and Indeterminacy', *Dialectica*, 67: 427–53. 161

Martínez, Manolo. 2015. 'Informationally-Connected Property Clusters, and Polymorphism', *Biology & Philosophy*, 30: 99–117. 161

Maturana, H. R., and F. J. Varela. 1980. *Autopoiesis and Cognition. The Realization of the Living*. Dordrecht: Reidel. 58

Miall, R. Christopher, and Daniel M. Wolpert. 1996. 'Forward Models for Physiological Motor Control', *Neural Networks*, 9: 1265–79. 184

Millikan, Ruth Garrett. 1984. *Language, Thought and Other Biological Categories*. Cambridge, MA: MIT Press. 16, 18, 21, 60, 78, 100, 156, 158, 169, 188

Millikan, Ruth Garrett. 1989. 'Biosemantics', *Journal of Philosophy*, 86: 281–97. 158, 202

Millikan, Ruth Garrett. 1990. 'Truth Rules, Hoverflies, and the Kripke-Wittgenstein Paradox', *Philosophical Review*, 99: 323–53. 158, 159

Millikan, Ruth Garrett. 1995. 'A Bet with Peacocke'. In Macdonald and Macdonald, eds., *Philosophy of Psychology: Debates on Psychological Explanation*. Oxford: Blackwell, 285–92. 158

Millikan, Ruth Garrett. 1996. 'On Swampkinds', *Mind & Language*, 11: 103–17. 22, 169

Millikan, Ruth Garrett. 2000. *On Clear and Confused Ideas*. Cambridge University Press. 13, 38, 77

Millikan, Ruth Garrett. 2004. *Varieties of Meaning*. London/Cambridge, MA: MIT Press. 158, 191, 192

Millikan, Ruth Garrett. 2009. 'Biosemantics'. In MacLaughlin, ed., *The Oxford Handbook of Philosophy of Mind*. Oxford University Press, 394–406. 159

Milner, A. D., and M. A. Goodale. 2006. *The Visual Brain in Action*. 2nd edn. Oxford University Press. 53

Moore, Michael T., and David M. Fresco. 2012. 'Depressive Realism: A Meta-Analytic Review', *Clinical Psychology Review*, 32: 496–509. 172

Mossio, Matteo, Cristian Saborido, and Alvaro Moreno. 2009. 'An Organizational Account of Biological Functions', *British Journal for the Philosophy of Science*, 60: 813–41. 58

Nagel, Ernest. 1977. 'Goal-Directed Processes in Biology', *Journal of Philosophy*, 74: 261–79. 54

Nanay, Bence. 2014. 'Teleosemantics without Etiology', *Philosophy of Science*, 81: 798–810. 73

Neander, Karen. 1995. 'Misrepresenting & Malfunctioning', *Philosophical Studies*, 79: 109–41. 159, 160

Neander, Karen. 2006. 'Content for Cognitive Science'. In Papineau and Macdonald, eds., *New Essays on Teleosemantics*. Oxford University Press. 160

Neander, Karen. 2017. *A Mark of the Mental: In Defense of Informational Teleosemantics*. London/Cambridge, MA: MIT Press. 27, 51, 139, 160, 161

Nieder, Andreas, and Stanislas Dehaene. 2009. 'Representation of Number in the Brain', *Annual Review of Neuroscience*, 32: 185–208. 98

Nisbett, Richard E., and Timothy D. Wilson. 1977. 'Telling More Than We Can Know: Verbal Reports on Mental Processes', *Psychological Review*, 84: 231. 28

O'Brien, Gerard, and Jon Opie. 2004. 'Notes Toward a Structuralist Theory of Mental Representation'. In Clapin, Staines and Slezak, eds., *Representation in Mind: New Approaches to Mental Representation*. Amsterdam: Elsevier, 1–20. 112, 138

O'Connor, Cailin. 2014. 'Evolving Perceptual Categories', *Philosophy of Science*, 81: 840–51. 216

O'Keefe, John, and Neil Burgess. 1996. 'Geometric Determinants of the Place Fields of Hippocampal Neurons', *Nature*, 381: 425–8. 113

O'Keefe, John, and Neil Burgess. 2005. 'Dual Phase and Rate Coding in Hippocampal Place Cells: Theoretical Significance and Relationship to Entorhinal Grid Cells', *Hippocampus*, 15: 853–66. 114

O'Keefe, John, and Lynn Nadel. 1978. *The Hippocampus as a Cognitive Map*. Oxford: Clarendon Press. 113

Ólafsdóttir, H. Freyja, Caswell Barry, Aman B. Saleem, Demis Hassabis, and Hugo J. Spiers. 2015. 'Hippocampal Place Cells Construct Reward Related Sequences through Unexplored Space', *Elife*, 4: e06063. 115

Papineau, David. 1987. *Reality and Representation*. Oxford: Blackwell. 16, 21

Papineau, David. 2003. 'Is Representation Rife?', *Ratio*, 16: 107–23. 16, 159

Papineau, David. 2016. 'Teleosemantics'. In Smith, ed., *How Biology Shapes Philosophy*. Cambridge University Press. 159, 170

Passingham, Richard. 2008. *What Is Special about the Human Brain?* Oxford University Press. 135

Peacocke, Christopher. 1992. *A Study of Concepts*. Cambridge, MA: MIT Press. 13, 155

Peacocke, Christopher. 1993. 'Externalist Explanation', *Proceedings of the Aristotelian Society*, 93: 203–30. 32, 172, 206

Pfeiffer, Brad E., and David J. Foster. 2013. 'Hippocampal Place-Cell Sequences Depict Future Paths to Remembered Goals', *Nature*, 497: 74–9. 115, 116

Piazza, Manuela, Veronique Izard, Philippe Pinel, Denis Le Bihan, and Stanislas Dehaene. 2004. 'Tuning Curves for Approximate Numerosity in the Human Intraparietal Sulcus', *Neuron*, 44: 547–55. 98

Pietroski, Paul. 1992. 'Intentionality and Teleological Error', *Pacific Philosophical Quarterly*, 73: 267–81. 154

Ponulak, Filip, and John J. Hopfield. 2013. 'Rapid, Parallel Path Planning by Propagating Wavefronts of Spiking Neural Activity', *Frontiers in Computational Neuroscience*, 7: 1–14. 115.

Pouget, A., P. Dayan, and R. S. Zemel. 2003. 'Inference and Computation with Population Codes', *Annual Review of Neuroscience*, 26: 381–410. 8

Pravosudov, Vladimir V., and Nicola S. Clayton. 2001. 'Effects of Demanding Foraging Conditions on Cache Retrieval Accuracy in Food-Caching Mountain Chickadees (Poecile Gambeli)', *Proceedings of the Royal Society of London B: Biological Sciences*, 268: 363–8. 50

Price, Carolyn. 2001. *Functions in Mind*. Oxford: Clarendon Press. 150, 159–60, 189

Putnam, Hilary. 1981. *Reason, Truth and History*. Cambridge University Press. 139

Ramsey, William. 1997. 'Do Connectionist Representations Earn Their Explanatory Keep?', *Mind & Language*, 12: 34–66. 205–6, 207

Ramsey, William. 2007. *Representation Reconsidered*. Cambridge University Press. 10, 30, 32, 118, 128, 206

Recanati, François. 2012. *Mental Files*. Oxford University Press. 13, 38

Redding, Gordon M., and Benjamin Wallace. 1997. 'Prism Adaptation during Target Pointing from Visible and Nonvisible Starting Locations', *Journal of Motor Behavior*, 29: 119–30. 53

Reid, Alliston K., and John E. R. Staddon. 1997. 'A Reader for the Cognitive Map', *Information Sciences*, 100: 217–28. 115

Reid, Alliston K., and John E. R. Staddon. 1998. 'A Dynamic Route Finder for the Cognitive Map', *Psychological Review*, 105: 585. 115

Rescorla, Michael. 2009a. 'Predication and Cartographic Representation', *Synthese*, 169: 175–200. 125

Rescorla, Michael. 2009b. 'Cognitive Maps and the Language of Thought', *British Journal for the Philosophy of Science*, 60: 377–407. 125

Rolls, E. T. 2015. 'Taste, Olfactory, and Food Reward Value Processing in the Brain', *Progress in Neurobiology*, 127–8: 64–90. 87

Rushworth, M. F. S., R. B. Mars, and C. Summerfield. 2009. 'General Mechanisms for Making Decisions?', *Current Opinion in Neurobiology*, 19: 75–83. 7

Rushworth, M. F., M. P. Noonan, E. D. Boorman, M. E. Walton, and T. E. Behrens. 2011. 'Frontal Cortex and Reward-Guided Learning and Decision-Making', *Neuron*, 70: 1054–69. 87

Ryder, D. 2004. 'Sinbad Neurosemantics: A Theory of Mental Representation', *Mind & Language*, 19: 211–40. 160–1, 186

Ryder, Dan. Forthcoming. *Models in the Brain: A Theory of Human Intentionality*. Oxford University Press. 186

Sainsbury, Mark, and Michael Tye. 2007. *Seven Puzzles of Thought: And How to Solve Them: An Originalist Theory of Concepts*. Oxford University Press. 13, 38

Salmon, Wesley. 1984. *Scientific Explanation and the Causal Structure of the World*. Princeton University Press. 88

Samsonovich, Alexei V., and Giorgio A. Ascoli. 2005. 'A Simple Neural Network Model of the Hippocampus Suggesting Its Pathfinding Role in Episodic Memory Retrieval', *Learning & Memory*, 12: 193–208. 115

Scheffler, Israel. 1959. 'Thoughts on Teleology', *British Journal for the Philosophy of Science*, IX: 265–84. 55

Schindler, Igor, Nichola J. Rice, Robert D. McIntosh, Yves Rossetti, Alain Vighetto, and A. David Milner. 2004. 'Automatic Avoidance of Obstacles Is a Dorsal Stream Function: Evidence from Optic Ataxia', *Nature Neuroscience*, 7: 779–84. 53, 54

Schlosser, Gerhard. 1998. 'Self-Re-Production and Functionality', *Synthese*, 116: 303–54. 58

Schulte, Peter. 2015. 'Perceptual Representations: A Teleosemantic Answer to the Breadth-of-Application Problem', *Biology & Philosophy*, 30: 119–36. 203

Searcy, William A., and Stephen Nowicki. 2005. *The Evolution of Animal Communication*. Princeton University Press. 18

Segal, G. 1991. 'Defence of a Reasonable Individualism', *Mind*, 100: 485–94. 35

Seyfarth, R. M., D. L. Cheney, and P. Marler. 1980. 'Vervet Monkey Alarm Calls: Semantic Communication in a Free-Ranging Primate', *Animal Behaviour*, 28: 1070–94. 119

Shagrir, Oron. 2001. 'Content, Computation and Externalism', *Mind*, 110: 369–400. 40

Shagrir, Oron. 2006. 'Why We View the Brain as a Computer', *Synthese*, 153: 393–416. 206

Shagrir, Oron. 2012. 'Structural Representations and the Brain', *British Journal for the Philosophy of Science*, 63: 519–45. 118

Shannon, Claude E. 1948. 'A Mathematical Theory of Communication', *Bell System Technical Journal*, 27: 379–423, 623–56. 12

Shea, Nicholas. 2007a. 'Content and Its Vehicles in Connectionist Systems', *Mind & Language*, 22: 246–69. 34, 216, 217

Shea, Nicholas. 2007b. 'Consumers Need Information: Supplementing Teleosemantics with an Input Condition', *Philosophy and Phenomenological Research*, 75: 404–35. 18, 43, 72, 159, 208, 209

Shea, Nicholas. 2007c. 'Representation in the Genome, and in Other Inheritance Systems', *Biology & Philosophy*, 22: 313–31. 19

Shea, Nicholas. 2009. 'Imitation as an Inheritance System', *Philosophical Transactions of the Royal Society B*, 364: 2429–43. 19

Shea, Nicholas. 2011a. 'Developmental Systems Theory Formulated as a Claim About Inherited Information', *Philosophy of Science*, 78: 60–82.

Shea, Nicholas. 2011b. 'What's Transmitted? Inherited Information', *Biology & Philosophy*, 26: 183–9. 19

Shea, Nicholas. 2011c. 'New Concepts Can Be Learned', *Biology & Philosophy*, 26: 129–39. 126

Shea, Nicholas. 2012a. 'Genetic Representation Explains the Cluster of Innateness-Related Properties', *Mind & Language*, 27: 466–93. 19

Shea, Nicholas. 2012b. 'New Thinking, Innateness and Inherited Representation', *Philosophical Transactions of the Royal Society B*, 367: 2234–44. 19

Shea, Nicholas. 2013a. 'Inherited Representations Are Read in Development', *British Journal for the Philosophy of Science*, 64: 1–31. 19

Shea, Nicholas. 2013b. 'Naturalising Representational Content', *Philosophy Compass*, 8: 496–509. 39

Shea, Nicholas. 2013c. 'Millikan's Isomorphism Requirement'. In Kingsbury, Ryder, and Williford, eds., *Millikan and Critics*. Oxford/Malden, MA: Wiley-Blackwell, 63–80. 14

Shea, Nicholas. 2014a. 'Exploited Isomorphism and Structural Representation', *Proceedings of the Aristotelian Society*, 64: 123–44. 127

Shea, Nicholas. 2014b. 'Neural Signaling of Probabilistic Vectors', *Philosophy of Science*, 81: 902–13. 106

Shea, Nicholas. 2014c. 'Reward Prediction Error Signals Are Meta-Representational', *Nous*, 48: 314–41. 218

Shea, Nicholas. 2015. 'Distinguishing Top-Down from Bottom-up Effects'. In Biggs, Matthen, and Stokes, eds., *Perception and Its Modalities*. Oxford University Press, 73–91. 138, 223

Shea, Nicholas. 2016. 'Representational Development Need Not Be Explicable-by-Content'. In Müller, ed., *Fundamental Issues of Artificial Intelligence*. Switzerland: Springer Synthese Library. 216

Shea, Nicholas, Ido Pen, and Tobias Uller. 2011. 'Three Epigenetic Information Channels and Their Different Roles in Evolution', *Journal of Evolutionary Biology*, 24: 1178–87. 19

Shea, Nicholas, Peter Godfrey-Smith, and Rosa Cao. 2017. 'Content in Simple Signalling Systems', *British Journal for the Philosophy of Science*, doi: 10.1093/bjps/axw036. 79, 219, 222

Shields, C. 2013. 'Aristotle'. In Zalta, ed., *The Stanford Encyclopedia of Philosophy* (Winter 2013 edn). http://plato.stanford.edu/archives/win2013/entries/aristotle 48.

Shigihara, Yoshihito, and Semir Zeki. 2013. 'Parallelism in the Brain's Visual Form System', *European Journal of Neuroscience*, 38: 3712–20. 104

Skyrms, Brian. 2010. *Signals: Evolution, Learning, & Information*. Oxford University Press. 57, 79, 190, 217, 219

Smith, Maurice A., and Reza Shadmehr. 2005. 'Intact Ability to Learn Internal Models of Arm Dynamics in Huntington's Disease but Not Cerebellar Degeneration', *Journal of Neurophysiology*, 93: 2809–21. 53

Smith, Michael. 1987. 'The Humean Theory of Motivation', *Mind*, 96: 36–61. 191

Sober, Elliott. 1994. *The Nature of Selection: Evolutionary Theory in Philosophical Focus*. Cambridge, MA: MIT Press. 150

Sommerhoff, G. 1950. *Analytical Biology*. London/New York: Oxford University Press. 54

Srinivasan, M., Shaowu Zhang, M. Lehrer, and T. Collett. 1996. 'Honeybee Navigation en route to the Goal: Visual Flight Control and Odometry', *Journal of Experimental Biology*, 199: 237–44. 72

Stegmann, Ulrich E. 2009. 'A Consumer-Based Teleosemantics for Animal Signals', *Philosophy of Science*, 76: 864–75. 96

Sterelny, Kim. 1995. 'Basic Minds', *Philosophical Perspectives*, 9: 251–70. 202

Sterelny, Kim. 2003. *Thought in a Hostile World*. Oxford: Blackwell. 189

Sterelny, Kim. 2015. 'Content, Control and Display: The Natural Origins of Content', *Philosophia*, 43: 549–64. 64

Stich, Stephen P. 1983. *Folk Psychology and Cognitive Science: The Case against Belief*. Cambridge, MA: MIT Press. 206

Stoljar, Daniel. 2001. 'The Conceivability Argument and Two Conceptions of the Physical', *Philosophical Perspectives*, 15: 393–413. 3

Suarez, Mauricio. 2003. 'Scientific Representation: Against Similarity and Isomorphism', *International Studies in the Philosophy of Science*, 17: 225–44. 112

Swoyer, Chris. 1991. 'Structural Representation and Surrogative Reasoning', *Synthese*, 87: 449–508. 118

Szalay, Máté S., István A. Kovács, Tamás Korcsmáros, Csaba Böde, and Péter Csermely. 2007. 'Stress-Induced Rearrangements of Cellular Networks: Consequences for Protection and Drug Design', *FEBS Letters*, 581: 3675–80. 214

Takahashi, Hideyuki. 1997. 'Hydrotropism: The Current State of Our Knowledge', *Journal of Plant Research*, 110: 163. 213

Thiele, Alexander, Karen R. Dobkins, and Thomas D. Albright. 2001. 'Neural Correlates of Chromatic Motion Perception', *Neuron*, 32: 351–8. 104

Thoroughman, Kurt A., and Reza Shadmehr. 2000. 'Learning of Action through Adaptive Combination of Motor Primitives', *Nature*, 407: 742–7. 53

Usher, Marius. 2001. 'A Statistical Referential Theory of Content: Using Information Theory to Account for Misrepresentation', *Mind & Language*, 16: 311–34. 12

Wagner, Elliott O. 2012. 'Deterministic Chaos and the Evolution of Meaning', *British Journal for the Philosophy of Science*, 63: 547–75. 220

Wagner, Elliott O. 2015. 'Conventional Semantic Meaning in Signalling Games with Conflicting Interests', *British Journal for the Philosophy of Science*, 66: 751–73. 220

Walsh, Denis. 2012. 'Mechanism and Purpose: A Case for Natural Teleology', *Studies in History and Philosophy of Science Part C: Studies in History and Philosophy of Biological and Biomedical Sciences*, 43: 173–81. 54

Whyte, J. 1990. 'Success Semantics', *Analysis*, 50: 149–57. 16

Williams, J. Robert G. 2016. 'Representational Scepticism: The Bubble Puzzle', *Philosophical Perspectives*, 30: 419–42. 14, 224

Williams, J. Robert G. 2018. 'Normative Reference Magnets', *Philosophical Review*. 14, 223, 127: 41–71

Wilson, Matthew A., and Bruce L. McNaughton. 1994. 'Reactivation of Hippocampal Ensemble Memories During Sleep', *Science*, 265: 676–9. 115

Wolpert, Daniel M., R. Chris Miall, and Mitsuo Kawato. 1998. 'Internal Models in the Cerebellum', *Trends in Cognitive Sciences*, 2: 338–47. 26, 27

Wolpert, Daniel M., J. Diedrichsen, and J. R. Flanagan. 2011. 'Principles of Sensorimotor Learning', *Nature Reviews Neuroscience*, 12: 739–51. 67

Wolpert, Daniel M., and Zoubin Ghahramani. 2000. 'Computational Principles of Movement Neuroscience', *Nature Neuroscience*, 3: 1212–17. 67, 183

Wolpert, Daniel M., and Michael S. Landy. 2012. 'Motor Control Is Decision-Making', *Current Opinion in Neurobiology*, 22: 996. 60

Wouters, Arno G. 1995. 'Viability Explanation', *Biology & Philosophy*, 10: 435–57. 58

Wouters, Arno G. 2007. 'Design Explanation: Determining the Constraints on What Can Be Alive', *Erkenntnis*, 67: 65–80. 58

Wright, Larry. 1973. 'Functions', *Philosophical Review*, 82: 139–68. 56–7

Xu, F., and E. S. Spelke. 2000. 'Large Number Discrimination in 6-Month-Old Infants', *Cognition*, 74: B1–B11. 98

Yang, Tianming, and Michael N. Shadlen. 2007. 'Probabilistic Reasoning by Neurons', *Nature*, 447: 1075–80. 25, 26

Zollman, K. J. S. 2011. 'Separating Directives and Assertions Using Simple Signaling Games', *Journal of Philosophy*, 108: 158–69. 190

Index